Geographical Information Retrieval in Textual Corpora

FOCUS SERIES

Series Editor Anne Ruas

Geographical Information Retrieval in Textual Corpora

Christian Sallaberry

iSTE

WILEY

First published 2013 in Great Britain and the United States by ISTE Ltd and John Wiley & Sons, Inc.

ISTE Ltd
27-37 St George's Road
London SW19 4EU
UK

www.iste.co.uk

John Wiley & Sons, Inc.
111 River Street
Hoboken, NJ 07030
USA

www.wiley.com

Library of Congress Control Number: 2013940049

British Library Cataloguing-in-Publication Data
A CIP record for this book is available from the British Library
ISSN: 2051-2481 (Print)
ISSN: 2051-249X (Online)
ISBN: 978-1-84821-596-2

MIX
Paper from responsible sources
FSC® C013604

Contents

Foreword

This very well-documented book addresses the field of geographic information extraction and retrieval from textual documents. Geographic information retrieval from documents is, indeed, a rapidly emerging subject, a trend fostered by the growing power of the Internet and the emerging possibilities of data dissemination. Information is processed from the identification of spatial and temporal features in textual documents, data indexing and manipulation of the relevance of identified items, multicriteria retrieval and an evaluation of query results by the development of several prototypes.

The author first introduces the principles of document retrieval and then illustrates the roles and importance of spatial and temporal information in textual documents. The addressed scientific challenges lie at the intersection of information retrieval techniques, natural language processing and qualitative spatial reasoning. The contributions presented address the development of spatial and temporal data models, geographic information extraction and analysis as well as symbolic annotations. Christian Sallaberry develops several of his recent contributions oriented around the development of spatial and temporal information indexing and textual document retrieval, these propositions being, by themselves, a worthwhile contribution of this monograph.

The book is usefully completed by a rich bibliographical study of current approaches focused on the modeling and retrieval of spatial and temporal information in textual documents and similarity measures developed so far in published literature. This allows Christian Sallaberry to develop a contribution in which the linguistic annotations, as well as the developed framework, enable us to identify, interpret and retrieve spatiotemporal information. This approach is typically qualitative in the sense that the spatial and temporal features identified in a corpus can be described from spatial and temporal relationships. These relationships play an important role in the derivation of spatial and temporal indexes and the execution of information retrieval processes, where spatial and temporal similarity measures allow us to trigger and rank query results.

The framework is completed by a multicriteria information retrieval approach. To develop and present his contribution, Christian Sallaberry introduces a useful literature review of spatiotemporal query homogenization. He introduces a spatial and temporal indexing approach based on the concepts of tiling and relevance scores, and different degrees of preference levels.

The conclusion of this book provides a broad perspective on the remaining scientific challenges. Several areas of research are discussed: integration of a domain-based ontology, modeling spatial relations in the interpretation of spatial features, generalization of these approaches in relation to the temporal and semantic dimensions, and semantic enrichment from annotations. All these domains are challenging and very attractive areas of research.

Overall, this book constitutes of a very well documented contribution, original and useful in a domain undergoing rapid development. The approach is original and brings a contribution to the field of geographic information extraction and retrieval from textual documents. It should raise wide interest for researchers in the fields of geographic and textual information processing as well as in developers of information and Web data processing systems. I hope it will generate many new vocations!

Dr Christophe CLARAMUNT
Professor in Computer Science
Naval Academy Research Institute
Lanveoc-Poulmic, June 2013

Acknowledgments

This book has its origins in my accreditation to direct research. Thus, my acknowledgments go first to Mauro Gaio, Professor at the UPPA[1], for his help in this long preparation for the accreditation to direct research. I would also like to thank my reviewers Mohand Boughanem, Professor at the Paul Sabatier University in Toulouse, Christophe Claramunt, Professor at the Naval Academy of Brest, and Ross Purves, Professor at the University of Zurich, for their expert reports and numerous pieces of advice that have enabled me to improve the original manuscript. Finally, thanks go to Marie-Aude Aufaure, Professor at the Ecole Centrale de Paris, Florence Le Ber, PhD Supervisor at the Ecole nationale du génie de l'eau et de l'environnement de Strasbourg, and Thierry Nodenot, Professor at the UPPA, who have also carefully examined the manuscript of my accreditation to direct research.

I will present the results of the work conducted as a team within the laboratory of LIUPPA[2]. Therefore, I would like to thank once again Mauro Gaio for associating me with his research works on natural language processing and reasoning aiming at spatial and temporal information marking and analyzing in bodies of text. My thanks go equally to my colleagues at the LIUPPA. I would like to mention, in particular, my colleagues Marie-Noëlle Bessagnet, Annig Lacayrelle and Albert Royer as well as the four doctoral students Pierre Laforcade, Julien Lesbegueries, VanTien NGuyen and Damien Palacio whose works I have been able to jointly accompany: they have offered me the possibility to share in fruitful collaborations that have contributed a lot

1 University of Pau and Pays de l'Adour: www.univ-pau.fr/.

2 Computer Science Laboratory of the University of Pau and Pays de l'Adour: liuppa.univ-pau.fr/.

to this research. I also address my gratitude to all other colleagues with whom I have had the chance of working within the context of different research projects. I would like to specifically thank my colleagues at the IRIT[3] institute of Toulouse, Guillaume Cabanac and Gilles Hubert, for their confidence and their pertinent proposals that have contributed a lot to these results.

Without all these meetings, the work presented in this book would not have seen the light of day.

3 Institut de recherche en informatique de Toulouse: www.irit.fr/.

Introduction

I.1. Geographic information retrieval

The work presented in this book lies within the field of geographic information retrieval (GIR). Information retrieval (IR) is finding documents which satisfy an information need from within a collection of documents generally stored on the Internet [MAN 08b]. GIR, first named and defined by Ray Larson [LAR 96], aims at retrieving documents which satisfy geographic characteristics: thus, the geographic zones featured in documents resulting from a GIR partially or entirely cover those expressed in the query. The series of GIR conferences[1] that began in 2004 [PUR 04] has heavily contributed to the development of GIR. GIR focuses on the spatial dimension in the first place and then, for textual documents, is extended by the thematic dimension conveyed by meaningful terms (other than spatial). We find the spatio-textual search dimension in the series of SSTD[2] conferences [JEN 01] beginning in 2005 [VAI 05], GeoCLEF[3] [GEY 05] beginning in 2005 [BUC 05] and GIS[4] [PIS 93] beginning in 2007 [LIE 07]. More notably, it is the series of RIAO[5] [ARS 85], GIR and GIS conferences that associated the temporal dimension with the spatial dimension and/or the thematic dimension: spatio-temporal-textual search, respectively, in 2004 [WID 04], in 2007 [MAR 07] and in 2010 [LIU 10].

Numerous research publications discuss these dimensions of GIR. We can name the books *Georeferencing* [HIL 06]; *The Geospatial Web* [SCH 07]; *Linguistique et recherche d'information, la problématique du temps* (*Linguistics and information retrieval, the temporal issue*) [BAT 11] or the theses "Toponym resolution by text"

1 http://www.geo.unizh.ch/~rsp/gir10/.

2 http://dblab.cs.ucr.edu/conferences/sstd01/.

3 http://ir.shef.ac.uk/geoclef/2005/.

4 http://www.informatik.uni-trier.de/~ley/db/conf/gis/.

5 http://www.informatik.uni-trier.de/~ley/db/conf/riao/.

[LEI 07]; "Geographic aware web text mining" [MAR 08a]; "Temporal information retrieval" [ALO 08]; "Geographic information retrieval: classification, disambiguation and modeling" [OVE 09], "Geographically constrained information retrieval" [AND 10], "Traitement automatique du langage pour l'indexation thématique et l'extraction d'informations temporelles" (*Natural Language Processing for the extraction and indexing of thematic and temporal information*) [KEV 11]. These works mainly target GIR in textual or multimedia documents consisting of a few lines to a number of pages available on the Web.

The work presented in this book focuses on digital libraries (DL) and, in particular, textual corpora as the application domain. We can refer to the Google Books[6] project with more than 10 million books digitized to date, the World Digital Library[7] with 6,142 digital objects at the moment, the Europeana[8] project with 10 million digitized objects to date or the Gallica[9] project of the National Library of France (BnF) with more than a million textual documents (books, periodicals, reviews and journals) digitized. Similar to many libraries and multimedia libraries, the MIDR[10] of Pau Pyrénées digitizes various kinds of documents (literary works, travelogues, newspapers, old geographical maps, lithographs, postcards, etc.), which have a common attribute of dealing with a small territory (the Pyrénées[11]) in a given period of history (mainly the 18th and 19th Centuries). This kind of document repository contains a great deal of references to history, geography, heritage; in other words to the territory [KER 11]. The objective of these different projects is to provide, to the widest audience, new means of accessing document repositories now available in digital formats. Thus, these projects implement processes of marking information, constructing indexes and querying by using these indexes.

The documents composing the corpus of MIDR are of particular importance in their richness in geographical indications relative to the Pyrenean territory. User categories such as "tourist", "student", "pedagogue", "scholar" and "librarian" have been identified by the staff of MIDR. These users intend to take advantage of the corpus by using an adapted information system capable, in particular, of offering search possibilities from the viewpoint of the territory represented by the corpus. As stated by Jihad Farhat and Luc Girard [FAR 04], document management systems (DMS) and search engines complement each other in order to support the activities of library users and professionals. We propose extending the functionalities of these systems through specific services dedicated to the processing of the spatial, temporal

6 http://books.google.fr/books/.

7 http://www.wdl.org.

8 http://www.europeana.eu/portal/.

9 http://gallica.bnf.fr/.

10 Médiathèque Intercommunale à Dimension Régionale de Pau Pyrénées – http://www.agglo-pau.fr/.

11 http://en.wikipedia.org/wiki/Pyrénées/.

and thematic dimensions of information. Thus, in comparison to content on the Web, we only consider document repositories such as those of MIDR that are stable (the content of a book does not change over the course of time) and homogeneous to allow thorough indexing relative to each of these three dimensions.

I.2. From spatial and temporal information indexing to multicriteria information retrieval

Literature relative to GIR in textual corpora presents the following challenges:

1) the recognition and interpretation of the spatial and temporal named entities;

2) the spatial and temporal indexation for purposes of IR;

3) the matching of document/query couples and the calculation of relevance scores dedicated to spatial IR on the one hand and temporal IR on the other hand;

4) the multicriteria IR combining the spatial, temporal and thematic dimensions;

5) the evaluation of such GIR systems.

In the laboratory of LIUPPA[12], within the T2I[13] team, the work corresponding to point 1, under the direction of Mauro Gaio [GAI 08], constitutes the basis of the work relative to points 2–5 [PAL 12a].

Thus, the recognition and determination of the spatial and temporal named entities [LEI 11, BAT 11] in textual documents is supported by two main classes of approaches. The first class relies on a set of rules, established by experts, allowing an interpreter to determine whether a term is a named entity or not. The second class is based on a manually annotated learning corpus allowing, after statistical processing, the automatic construction of rules for the discovery of named entities which can be applied to larger corpora. In accordance with the first class of rules, our work team proposes a set of manually built rules dedicated to the expression of space and time in a corpus composed of travelogues: these rules allow the marking as well as the first symbolic interpretation of the detected entities (classification followed by an analysis of a last associated spatial or temporal relation). Following this interpretation, we have distinguished absolute entities such as "the City of Pau" and "the year 2000" from the so-called relative entities such as "the surroundings of the City of Pau" and "at the beginning of the year 2000". Let us recall that we are only

12 Laboratoire informatique de l'université de Pau et des Pays de l'Adour: liuppa.univ-pau.fr/.

13 Processing of spatial, temporal and thematic information for the adaptation of contextual and user interaction (Traitement des Informations spatiales, temporelles et thématiques pour l'adaptation de l'Interaction au contexte et à l'utilisateur): http://liuppa.univ-pau.fr/live/EquipesdeRecherche/Equipe_T2I/.

dealing with the textual contents of documents, regardless of their structure or associated meta-descriptions.

The indexation associates a numerical interpretation (geometry, calendar period) with the detected spatial and temporal entities in the texts. The organization of the indexes can, for example, dissociate completely the references from space and theme in the independent indexes or, on the other hand, can combine these two dimensions in specific structures stored in one and the same index [VAI 05]. As seen in Clough *et al.* [CLO 06], we have chosen to work on independent spatial, temporal and thematic indexes. The algorithms that interpret the symbolic representation of entities take the absolute and relative aspects of their description into account. The resulting numerical representation corresponds to the outcome of a search in such resources as gazetteers in the case of absolute entities, for example.

The matching and calculation of the relevance scores have equally been the subject of numerous propositions in spatial IR [AND 10] as well as in temporal IR [ALO 08, BAT 11]. As for the majority of these propositions, we have developed a spatial and a temporal IR supported by ad hoc formulas adapted to our corpus.

The combination of the spatial, temporal and/or thematic dimensions in GIR is generally implemented using filtering approaches [VAI 05, LIE 07]. For a greater power of expression in the querying process, we have introduced requirement and preference operators that can be associated with each search criterion. Taking into account the different levels of requirements as expressed in the query, we have developed a method of aggregation of the results coming from different IR systems (IRSs). This method is inspired by the aggregation approaches established in decision support systems [MAR 99] as well as in multicriteria information retrieval systems [FAR 08].

The implementation of the first GIR prototypes emphasizes the necessity to evaluate such systems [CAR 11, MAN 11]. However, with the exception of such campaigns as TEMPEVAL [VER 09] devoted to time and GEOCLEF [GEY 05] devoted to space and theme, there are, to our knowledge, no evaluation frameworks of GIR systems that combine the spatial, temporal and thematic dimensions of information. We have therefore proposed an experimental framework devoted to this type of evaluation. We have established a testing collection as well as an experimentation protocol which we implement for the evaluation of our prototypes.

To deal with these different lines of research, the book is organized into two main chapters (Chapters 2 and 3). Chapter 2 details the indexation and the retrieval of spatial and temporal information in textual corpora. We deal with spatial IR on the one hand and temporal IR on the other hand. Chapter 3 discusses at the handling of spatial and temporal indexes obtained earlier in the context of multicriteria information retrieval. We broadly discuss the GIR here because this is an IR that combines the spatial, temporal and thematic search criteria.

The indexing of spatial and temporal information in textual documents constitutes the basis of this work. In this indexing part, the quality of the recognition and interpretation of spatial and temporal named entities is paramount. The following processes use the results of this indexation for the purposes of separate spatial IR, separate temporal IR or multicriteria IR combining the three geographic dimensions.

Indexing of spatial and temporal information in textual documents

In Chapter 2, we first look at the modeling of the spatial and temporal information in the context of specialized information retrieval devoted to non-structured textual corpora (section 2.3.3). We propose spatial and temporal core models [LES 06, GAI 08] (section 2.4.2) devoted to such information interpretation (section 2.4.3) and representation in the indexes in order to implement matching calculations in the research phase. We design and experiment a first method of extraction and indexation of spatial information (section 2.4.4) based on our core model and a specific semantic processing [LES 06]. We adopt a similar approach in order to propose a method of extraction and indexation (section 2.4.4) of temporal information based on our core mode and a specific semantic processing [LEP 07].

Retrieval of spatial and temporal information in textual documents

In Chapter 2, we also describe the IR approaches implemented in the systems devoted to spatial and temporal information (sections 2.3.1 and 2.3.6). We propose a method of spatial information retrieval (section 2.4.5.1) using functions of geographic information systems (GISs) in order to calculate geo-referenced representations of spatial entities and implement spatial relevance calculations [SAL 07a]. Using a similar approach, we propose a method of dedicated temporal information retrieval [LEP 07] (section 2.4.5.2).

Generalization of data representations for multicriteria information retrieval

In Chapter 3, we deal with each dimension of the geographic information in a specific way and then combine them in IR scenarios. To avoid possible biases, it is important, before any combination, to standardize the representation of the data as well as the approaches of processing data relative to the different dimensions (section 3.3.1). We propose a generic approach comparable to the generalization by truncation or lemmatization of terms in classic approaches to IR. Thus, from indexed representations of spatial and temporal information, we build higher-level indexes (section 3.4.1) appropriate for the implementation of proven IR models [PAL 10c, SAL 11].

Multicriteria information retrieval

In Chapter 3, we also discuss the multicriteria information retrieval approaches (section 3.3.2). We propose submitting each search criterion to the appropriate IR system of spatial, temporal or thematic dimension (section 3.4.3). It should be noted that, for the thematic dimension, we limit ourselves to the approaches implemented for terms in classic IR. We offer several approaches for the combination of results from different indexes and IR systems. We also propose, according to the type of user involved, new operators with the aim of associating a higher level of expressiveness with each criterion of the query and, consequently, improving the quality of the results [PAL 10b, PAL 10c, PAL 11, PAL 12a].

I.3. Organization of the book

This book is divided into the following chapters:

– Chapter 1 presents the positioning of the work in the field of GIR.

– Chapter 2 is devoted to spatial and temporal information in textual documents. It describes our propositions relative to indexation and to spatial and temporal information retrieval in textual documents.

– Chapter 3 deals with the generalization of data representation with the aim of preparing the combination of results from multi-dimensional (spatial, temporal and thematic) and multicriteria information retrieval. This chapter describes our propositions for multi-dimensional and multicriteria information retrieval.

– Chapter 4 is devoted to the first overview followed by a presentation of a set of perspectives as extensions of this work in the field of GIR.

1

Access by Geographic Content to Textual Corpora: What Orientations?

1.1. Introduction

The volume of digital corpora is always on the rise and the retrieval of relevant documents is an increasingly delicate task. The ambiguity of natural language terms contributes to this difficulty in the automatic interpretation of the expression of the need for information as well as in the automatic evaluation of the correspondence between documents and needs. The multiple meanings of the terms and their numerous uses in varied contexts make delicate, indeed, the task of information retrieval. Our working hypothesis therefore consists of distinguishing the spatial, temporal and thematic dimensions in order to implement dedicated approaches in the processes of indexing and information retrieval (IR). The objective is to contribute to a better content analysis of textual corpora as well as to a better grasp of the search criteria expressed in a query. Let us recall that we are studying textual corpora with "territorial" denotations, digitized, to which processes of character recognition have been applied but whose logical structure has not been conserved.

This chapter is organized as follows. Section 1.2 presents the general context related to geographic information retrieval (GIR). Section 1.3 introduces privileged fields of research as well as the position of our study. Section 1.4 gives a rough sketch of our research approach in the construction of spatial, temporal and multicriteria search engines.

1.2. Access by geographic content to textual corpora

The study concerning the processing of information in text is mainly detailed in theses [BAZ 05, LES 07, PAL 10a, KER 11]. Following a number of reminders related to document retrieval and textual corpora, we will describe the characteristics of corpora with "territorial" denotations and their uses. This category of corpora will constitute the field of experimentation for our propositions.

1.2.1. *Document retrieval and textual corpora*

Document retrieval or *information retrieval* [BAE 99, BOU 08] is traditionally defined as a set of techniques allowing us to select, from a collection of documents, information that is likely to meet the needs of the user.

A *collection of documents* (document repository or corpus) is the information accessible via the document retrieval system (or information retrieval system, IRS). It consists of *documents*, unit elements. *Textual documents* are represented by a set of descriptors (terms, for example) stored in files of descriptive instructions (metadata) or indexes whose structure can be more complex [BES 04]. However, the notion of *document* in itself is vague. Generally defined by its container (e.g. a book, the physical object that contains the text), it often varies and the expected result of a query may not be an entire book but one or more particularly relevant fragments. This is indeed the reason why we use the expression "document unit" or "document fragment" to define the unit of text returned to the user [BAZ 05].

Finally, a *query* corresponds to the expression of the information needs of the user. It constitutes the input parameter to the retrieval system and is expressed in a query language that is often simple: a choice of keywords and logical operators, for instance. Nevertheless, other languages are presented in literature: natural language, graphical language, etc. [GOK 09].

1.2.2. *Textual corpora with "territorial" denotations*

A textual corpus with "territorial" denotations is composed of travelogues, stories, newspapers, novels, poems, etc. These documents describe/discuss a territory. As detailed in [KER 11], the territorial dimension is symbolized in textual documents by a significant frequency of toponyms, outlined facts or described observations. Toponyms denote, for example, streams, cities and buildings. The facts describe, for example, political or sport-related events as well as various other events. The observations refer to architecture, botany, geology, agriculture, etc. These categories of information are, in a general way, linked to a location or a period of time.

– *Territory:* The *Longman* Dictionary defines the term territory as "an area for which one person or branch of an organization is responsible". Kergosien [KER 11] presents a consistent overview of the notion of territory. Among the different definitions proposed, we will retain the following two [KER 11, p. 70]: "A globally accepted definition in geography describes territory as a space on which an authority is exercised and is limited by political and administrative borders. This definition is subject to debate, however the notion of territory generally integrates a geographic space composed of places (spatial component) as well as relations with different subjects (thematic component) and/or references to a period (temporal component)". It also describes a second point of view, that of geomatics. "Geomatics is the scientific field hovering between geography and computer science which mainly deals with problems of storage, processing and diffusion of geographic information. The characterization of geographic information in a particular territory is defined in the form of geographic entities (GEs) composed of spatial (SEs), temporal (TEs) and thematic entities. It should be noted that each one of these entities is not always specified or can be implicit". Kergosien [KER 11] proposes an approach of ontology construction as a tool for the structured representation of a territory but also as a support to IR and to the browsing of document repositories.

– *Examples of corpora:* Territory is at the heart of numerous types of corpora. We can quote, for example, the French-speaking corpus of archives, mainly composed of texts, maps and lithographies related to the city of Saint-Étienne and to its river Furan[1]; the multi-lingual corpus (German and French) of the Swiss Alpine Club[2], composed of reports, accounts, essays and thoughts under the theme of mountaineering; tourist guides such as the different ranges of *Lonely Planet*[3] books or of the *Michelin guide*[4]; and the equally numerous hiking guides[5] and other travel blogs[6].

These corpora have the principal characteristic of containing a very large number of place names (spatial named entities will be defined further on); the places referred to in such a way generally have a fine level of detail in a relatively confined space (a river, a city and mountain range, for example). The *Geotopia*[7] and *Text+Berg digital*[8] projects are good examples of this. The objective of the first is to experiment with geo-referencing techniques in order to help organize, transmit, share and interpret archival data [JOL 11]. The second aims to digitize and promote a corpus of alpine literature [VOL 10].

– *The corpus of MIDR:* MIDR[9], from a perspective of cultural heritage promotion, has digitized and implemented the optical character recognition of its heritage document repository with the aim of indexing it into a document retrieval system. This way, the digitized documents can benefit from a renewed visibility and be exploited

1 http://umrisig.wordpress.com/les-projets/projet-geotopia.

2 http://www.textberg.ch.

3 http://shop.lonelyplanet.com.

4 http://voyage.viamichelin.fr.

5 http://www.ffrandonnee.fr/boutique/le-catalogue-des-topo-guides.aspx.

6 http://www.blogs-de-voyage.fr.

7 http://umrisig.wordpress.com/les-projets/projet-geotopia.

8 http://www.textberg.ch.

9 Médiathèque Intercommunale à Dimension Régionale de Pau Pyrénées – http://www.agglo-pau.fr.

by a larger public. It should be noted that this digitalization, keeping in mind the cost of the operation, has been carried out by a provider without the correction of errors and the recovery of the documents' structure, with the exception of their division into paragraphs.

Let us recall that this corpus is composed of documents of different types (literary studies, travelogues, newspapers, old geographic maps, lithographies, postcards, etc.), which have the common denominator of dealing with the Pyrénées territory in the 18th and 19th Centuries. A preliminary study of the corpus has revealed a predominant geographic connotation in the documents, as much in the literary studies dealing with travelogues as in the local periodicals whose articles relate to information about the territory. An experimentation has allowed us, for example, to extract almost 10,000 spatial named entities from 10 books within the corpus (i.e. 600,000 terms).

Indeed, a large amount of information makes reference to places, spatial indications as well as descriptions of landscape, temporal indicators and dates, implying a significant importance of these documents for the geographic aspect. Let us consider, as an example, travelogues (see the excerpt in Figure 1.1). The authors of these pieces of study use, most of the time, an identical structure: the text is divided into sections describing a portion of their travel. Each portion can consist of the description of an itinerary, a stage, a point of view, an observation, an event, etc.

Journal of James David Forbes (excerpt) :

12 July. Sunday morning. This morning I left for Argeles. I left at 6:55. The firs start at 8:10. The rhododendrons at 9 o'clock. Shady saxifrage. Blue campanula. Saxifrage. 9:40 quite abundant spring Pass of Torte 9:50. The pass of Torte is in the slate. Very fine view on each side and, to the right, the peaks of Gourette. I descended in a valley and after passing a second cape I saw the beautiful valley of Azun spreading below in front of me and join the valley of Argelès in the distance. I rested and ate my provisions at Marsous. Argelès at about 6 o'clock. I believe today to have walked seven or eight hours; it is a quite tiring route but very beautiful. The peaks are very steep and the valley of Azun is very beautiful, just like the one coming from Eaux-Bonnes. In sum, this is quite worthy of Switzerland. The surroundings of Argelès are very beautiful. It is an alluvial land full with quite a surprising number of granite blocks, some of enormous size. They must have come down the valley of Azun, which is also plentiful in them. The size of these rocks is such that one could easily think they are there in situ. Sunday. Hôtel de Commerce very good. I ate at Eaux-Bonnes at the Hôtel de France: excellent.

13 July. I quit Argelès after breakfast and I went up the valley by walking until Luz, four or five leagues. Very pleasant and sometimes striking. The most beautiful is at the outskirts of Argelès. It is almost a paradise. I was tired from my walk yesterday and I was delighted to find myself at the Hôtel des Pyrénées of Luz, which, notwithstanding an unpromising exterior, had so many amenities inside that I immediately made it my headquarters. At night, I walked until St-Sauveur. At 9 o'clock, I wrote to Jane.

space
time
theme – botany
theme – minerals

Figure 1.1. *Document excerpt – The Travel to the Pyrénées, David James Forbes, CAIRN Editions (1835)*

Figure 1.1 represents two paragraphs from the travel journal of James David Forbes. In it we can find toponyms such as, for example, the "pass of Torre" whose toponymical reference name is "Torre". We can also observe temporal references such as "12 July" and thematic such as "granite" (to be considered, for example, from a mineral point of view) or "firs" (of particular interest from a botanical point of view). Let us also note the varying levels of the complexity of information: we refer to "Argelès" as a simple spatial information whereas "outskirts of Argelès" is a complex spatial piece of information that evokes a relation of adjacency whose interpretation necessitates an additional analysis of the text.

– *Geographic information:* The central element of the corpus being the geographic information, let us review a definition coming from geomatics: illustrated in Figure 1.2, it considers geographic information to be as a molecule not only composed of a spatial component, but also of a temporal component and a thematic component, or *phenomenon* [USE 96, GAI 01]. For example, the text "musical instruments in the vicinity of Laruns at the beginning of the 19th Century" fully describes this geographic molecule, with "musical instruments" corresponding to the thematic component. Let us note that some components might nonetheless be absent.

In the geography markup language (GML) specification[10] and the research on databases [LE 04], we can see the appearance of the notion of temporality. Thus, it is possible to associate a piece of geographic information with one or more geo-referenced representations, valid at a certain moment in history [GAL 01]. For instance, a city or a forest has a variable spatial definition over time which can be a creation, a disappearance, an expansion or a reduction. Finally, a phenomenon is often associated with it: subject of research, for example, a regional pollution at a given period of time [PAL 10a].

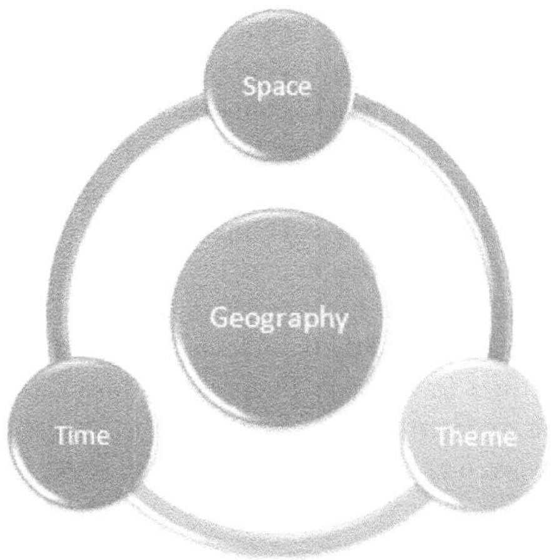

Figure 1.2. *Spatial, temporal and thematic dimensions of geographic information*

10 http://www.opengeospatial.org/standards/gml.

1.2.3. *Access to textual content*

A study conducted on IR tasks led by students has revealed that the three main categories "of search criteria" are of bibliographical (people), chronological (periods) and spatial (toponyms) types [MAN 09]. Many other studies show a considerable proportion of references to places in the search criteria of users: for the Excite [SAN 04], AOL [GAN 08] and Yahoo [JON 08] engines, this proportion varies between 12.7 and 18.6%. Moreover, 79.5% of these queries contain toponyms [SAN 04].

In the context of digital libraries (DLs), the interfaces of IR and navigation in the resulting documents are by default composed of a subject (themes) and chronological (see the Google Books, Europeana and Gallica projects) or subject, chronological and spatial (see the Bibliothèque Numérique Mondiale project) dimensions. Here, the IR process implements advanced document management tools. These document management systems are based on metadata composed of descriptive instructions or full-text indexes in which geographical information, toponyms among others, are exploited in the same way as all the other terms.

Concerning the corpus of MIDR, a number of categories of use could be studied.

A qualitative study of the activities of librarians in the case of event-preparation scenarios has allowed us to highlight IR approaches which prioritize, in order of importance, the categories of bibliography (people), subject (themes), chronology (periods) and place (toponyms). The usage scenarios of a tourist generally prioritize the current location of the tourist or the intended place to visit in order to later focus on the subject (themes), bibliography (people) or chronology (periods). We thus distinguish three categories of users potentially involved in IR composed of geographic criteria. Their basic knowledge is, *a priori*, decreasing. The first category includes scholars, for instance historians, who wish to find precise information related to a place or a date. It also encompasses librarians, for example, whose purpose is the improvement of document annotation or the preparation of exhibitions. The second category includes the inhabitants of a region who wish to know more about it. It also affects teachers and their students, for example, who want to discover the itinerary described in a travelogue. Finally, the third category includes tourists, for example, who wish to determine the activities, the monuments or other points of interest accessible in a given zone ("the canyons to the south of Laruns", "the springs around Pau", etc.). It also involves every person wanting to find information from spatial and temporal criteria.

We have highlighted the significant presence of geographic criteria in the IR scenarios applied to Web content and DLs. Nevertheless, the usual search engines do not allow us to take into account the particularities of spatial and temporal information. Indeed, they are limited to the search terms (keywords) entered by the users in their query. For instance, if we wish to find documents related to events associated with the south of Pau, the search engine will target the terms "south" and "Pau". However, a document referring to "Jurançon", which is a bordering commune to that of Pau and situated to its south, should also be returned. Similarly, for temporal information, if we wish to find documents describing events related to the 19th Century, the search engine should not only return the documents which contain "19th Century" but also those which contain "1801", "1802", etc.

Finally, an experienced user interested in documents related to the "Pyrénées mountains but not those of Gavarnie, in the 19th Century, if possible, unrelated to ascents" must be able to depict this type of information need and navigate in the set of resulting text units (paragraphs). To satisfy such needs, the construction of precise indexes adapted to each type of information (spatial, temporal and thematic) seems necessary. The aim is thus to improve GIR by combining the results obtained from devoted spatial and temporal processes as well as from classic IR strategies, employed generally for thematic criteria.

1.3. Reinforcement of GIR by contributions from NLP, reasoning and multicriteria IR

If we consider the association for computing machinery (ACM) classification[11], our study is related to section *H3 INFORMATION STORAGE AND RETRIEVAL* and, in particular, to subsections: *H.3.1 Content Analysis and Indexing*, *H.3.3 Information Search and Retrieval* and *H.3.7 Digital Libraries*. It concerns IR and, in particular, GIR in textual document repositories.

However, as we have already shown, our field of research is distinct from classic IR on a large number of points. We are interested in stable textual document repositories (*a priori*, no update of a given document of the repository) as well as those which are homogeneous in their style of expression (such as *travelogue*, *walk itinerary* and *tourist guide*). This particularity enables a thorough processing, on the one hand, for *back-office* indexing and specific usage scenarios, on the other. Concerning indexing, natural language processing (NLP) supports the targeted extraction and analysis of spatial and temporal information, while qualitative reasoning completes this analysis and supports the interpretation of this information as well as that of associated relations. Thus, Figure 1.3 positions our study concerning GIR at a cross-road between IR, NLP and qualitative reasoning. This can involve specialized IR dedicated to vocabulary proper to the expression of space and time. We propose an active parsing of the textual document, in other words a targeted search of expected elements of information in the text in order to build the corresponding spatial and temporal meaning of the speech.

11 ACM Computing Classification System – http://dl.acm.org.

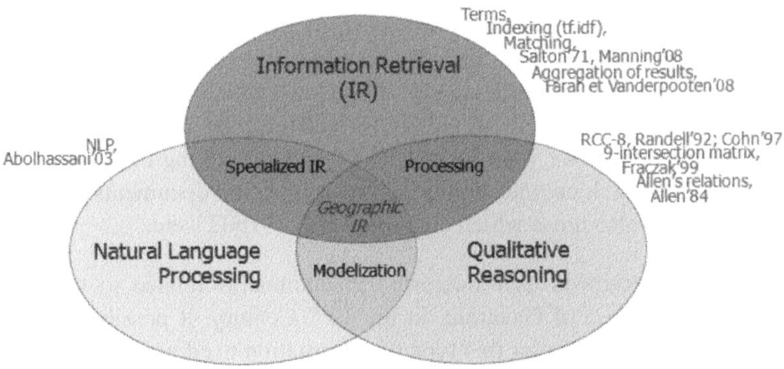

Figure 1.3. *Our GIR study and the main research fields of interest involved – extended extract of Julien Lesbegueries [LES 07]*

In Figure 1.3, we also show the necessity of having spatial and temporal models for the description of information extracted from texts. A number of processes, supported by geographic information systems (GIS), for example, dedicated to spatial and temporal information will allow us to calculate numeric representations (geometries, calendar periods) corresponding to the information described in the models.

Before describing our approach, which combines models and processes drawn from these different fields for the construction of a multicriteria search engine, we present the notions of IR, GIR, NLP and qualitative reasoning, some of which have already been briefly introduced.

Manning *et al.* [MAN 08b] define IR in two main processes with, on the one hand, techniques of document indexing and retrieval mechanisms, on the other. Thus (see section 1.2), the documents are represented by a set of descriptors (terms, for example) organized in an index. The user's need (query) is dealt with in a similar way: it is also represented by a set of descriptors. Afterwards, matching mechanisms compare the descriptors of the query with those contained in the index in order to build a list of relevant documents.

GIR differs from IR in its explicit recognition and modelization of space and time in the frame of indexing processes and IR [LEI 07, ALO 08]. In a GIR system, not only the key terms are indexed, but also the "spatial terms" with the corresponding geo-localizations called *spatial footprints* and the "temporal terms" with the corresponding intervals of time called *time stamps*. IR is, in this case, completed by the comparison of the spatial footprint or the time stamp of a query with the spatial footprints or time stamps extracted from the documents. In general, the comparison is based on the intersection sizes of these footprints and stamps.

The recognition of "spatial" and "temporal" terms in texts is supported by techniques of named entity recognition (NER). NER, detailed in [CHI 98], consists of the retrieval of textual objects called named entities (in other words, proper nouns, expressions of time and numeric expressions) which can be categorized into classes, such as people's names, names of organizations or businesses, place names, quantities, distances, values, dates, acronyms and abbreviations.

Moreover, the studies of linguistic analysis have found applications in the world of IR: for instance, the concept of *target/site* described by Vandeloise [VAN 86] and that of *concrete entity/reference* described by Borillo [BOR 98] shows the particular way humans describe spatial information when it comes to writing. It is more and more common to see techniques of linguistic analysis being associated with techniques of statistical analysis [DEL 04]: for example, the detection of named entities in a text uses morpho-syntactical processes of linguistic analysis [MIK 99]. Thus, the tools of NLP support a fine analysis based on the interpretation of the semantics contained in the textual documents. They contribute to NER and to the extraction of noun phrases which contain these entities. For example, "the south of Pau" is the phrase evoking the spatial named entity *Pau*, and "in the beginning of the 18th Century" is the phrase evoking the temporal named entity 18th Century.

Qualitative spatial reasoning (*QSR*) and qualitative temporal reasoning (*QTR*) complete the study of language by proposing reasoning processes for the acquisition of additional knowledge. The importance given to the qualitative aspects of spatial information stems from ancient Greece, as Kowalski *et al.* [KOW 07] recall. More recently, the study carried out by Allen [ALL 91] focuses on the temporal reasoning for a qualitative representation. Propositions for QSR have then adapted this study by taking into account the specificities and the bigger complexity of spatial information. Cohn [COH 96] and then later Cohn and Hazarika [COH 01] show the state of the art of QSR and, in particular, classify the spatial relations (see section 2.3.4). In the context of GIR, a reinvestment of the study related to QSR and QTR targets the interpretation of spatial and temporal noun phrases, respectively.

1.4. Toward the construction of a multicriteria IR engine

Accessing the content of textual documents via an IR approach integrating the spatial, temporal and thematic dimensions (Figure 1.2) is the main challenge of this study. Its objective is the construction of an IR engine combining these three dimensions.

1.4.1. *Challenges, hypotheses and research objectives*

1) *Challenges:* As we mentioned previously, the ratio of queries integrating spatial criteria, for example, varies between 12.7 and 18.6% according to Excite [SAN 04], AOL [GAN 08] and Yahoo [JON 08]. Although well tried today, the classic approaches to IR are limited in the case of geographic search criteria [LIE 09]. It remains true that what interests humans most often is the theme. However, taking into account the thematic dimension and more importantly the semantic aspects it holds is a very difficult task. Indeed, the current tools of IR are efficient but limited to terms. Our first objective, therefore, is to target the spatial and temporal dimensions as privileged entry points in the texts. The aim is, thus, to complete classic IRSs by specific services dedicated to the spatial and temporal aspects. The first challenge is therefore: *What models of representation and retrieval of spatial and temporal information should be proposed for the access by geographic content to textual corpora?*

It is then a question of combining a classic IRS with spatial and temporal IRSs. The heterogeneity of the models of representation and those of the corresponding IR does not allow us to directly consider the combination of such systems. The second challenge is thus: *Which core model of representation and retrieval of information should be proposed in order to prepare for the combination?*

Finally, the combination has to be based on aggregation operators adapted to the geographic context. Let us not forget the need for a power of expression highlighted earlier. This need is partially satisfied by the operators proper to each dimension. Nonetheless, it has to be completed with a finer formulation of each criterion: is it mandatory, is it associated with a level of preference, is it a rejection criterion? The third challenge is thus: *What advanced aggregation operators should be proposed and implemented for a multicriteria IR combining many IRSs?*

2) *Hypotheses:* We issue a study hypothesis for each of the challenges evoked. Concerning the implementation of specific process flows dedicated to spatial and temporal information, our study hypotheses are the following:

– A dedicated spatial IRS gives better results than a classic thematic IRS for IR composed of only spatial criteria.

– A dedicated temporal IRS gives better results than a classic thematic IRS for IR composed of only temporal criteria.

– A "rough" coupling of spatial, temporal and thematic IRSs gives better results than a classic thematic IRS for multicriteria IR despite the numerous possible biases linked, for example, to the heterogeneity of value domains with manipulated scores.

Finally, the implementation of a multicriteria IRS combining the results from spatial, temporal and thematic IRSs arises the following study hypotheses:

– The generalization of data representation is adapted to the spatial and temporal information.

– The generalization associated with a well-tried IR model does not imply a loss of quality with respect to the initial dedicated IRSs.

– A "classic" coupling (an arithmetic mean, for example) of spatial, temporal and thematic IRSs gives better results than a single thematic IRS or a two-by-two pairing of spatial, temporal and thematic IRSs.

– An "advanced" coupling (offering greater power of expression to the user) of spatial, temporal and thematic IRSs gives better results than a "classic" coupling.

3) *Objectives:* The research objectives corresponding to the formulated challenges mainly target the design and implementation of:

– models supporting the symbolic and numeric representation of spatial and temporal data;

– process flows of spatial and temporal information extraction, interpretation and indexing;

– spatial and temporal IR models;

– a model of generalization of the representation of indexed data;

– a model of multicriteria IR.

1.4.2. *Approach*

Our approach is divided into two main steps. First (Figure 1.4), under the direction of Gaio [GAI 08], we have implemented process flows dedicated to the automatic recognition of spatial and temporal named entities. We consider that, regardless of the territory, it is possible to clutch to a single or a set of named entities as characteristic elements of the content of a corpus of documents. The processes, successively lexical, morpho-syntactical and semantic consist, respectively, of extracting, validating and interpreting spatial and temporal named entities for the purposes of annotation and/or indexing. We propose a spatial and a temporal model dedicated to the needs of marking and representing the processed information in each of these steps. We obtain, in particular, spatial and temporal indexes in compliance with these models.

Texts annotated in such a way mainly target experts: librarians or archivists, for example, who need these results in order to create descriptive instructions (metadata) in a semi-automatic way. In addition, the indexes represent the content of document repositories and support IR scenarios as well as more sophisticated approaches for the discovery of territory targeting, for example pupils, tourists or scholars. We propose dedicated models of spatial and temporal IR.

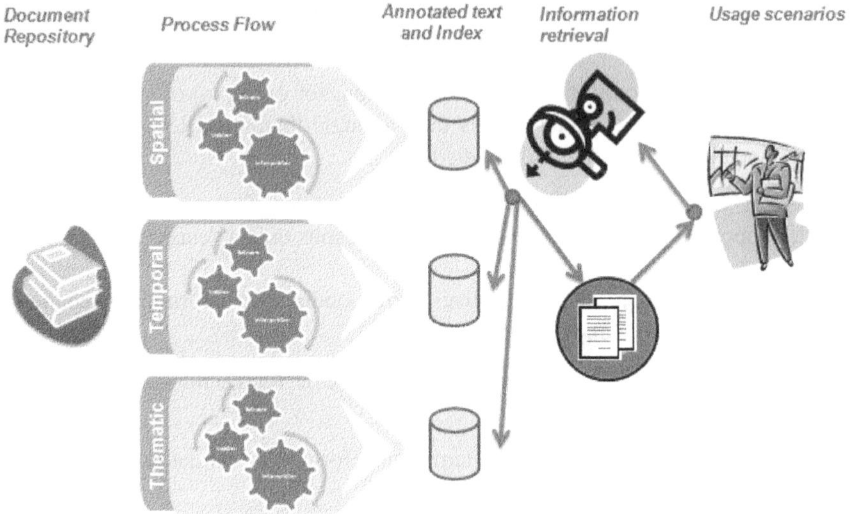

Figure 1.4. *Spatial and temporal process flows (extraction, validation and interpretation)*

Second (Figure 1.5), in continuity with the first step, we have implemented a multi-dimensional and a multicriteria IR model. We propose submitting each criterion of a query to the IRS dedicated to the corresponding spatial, temporal and thematic dimensions, followed by the combination of the results. Before any combination, however, in order to avoid possible biases, we have chosen to generalize the representation of the data corresponding to each dimension. This generalization necessitates first and foremost the segmentation of the space (respectively the period) covered by the document corpus to be indexed: we call this spatial (respectively temporal) tiling, or splitting. This results in generalized indexes (Figure 1.5). We then proceed to a *projection* in which every intersection between a tile of the generalized index and an object of the initial index increases the weight of the tile. We propose regular, administrative and calendar tilings with tiles of various size.

Thus, this tiling approach, comparable to the generalization by truncation or lemmatization of terms in classic IR approaches, allows us the implementation of well-tried IR models for each of the geographic dimensions. We compare, for example, vectorial IR to the ad hoc IR models developed for each dimension. Losses in precision and recall are of course induced by the generalization. Nevertheless, the integration of tile reference frequencies in the calculation of weight and relevance scores delivers gains that we have quantified (see section 3.5.2, Chapter 3).

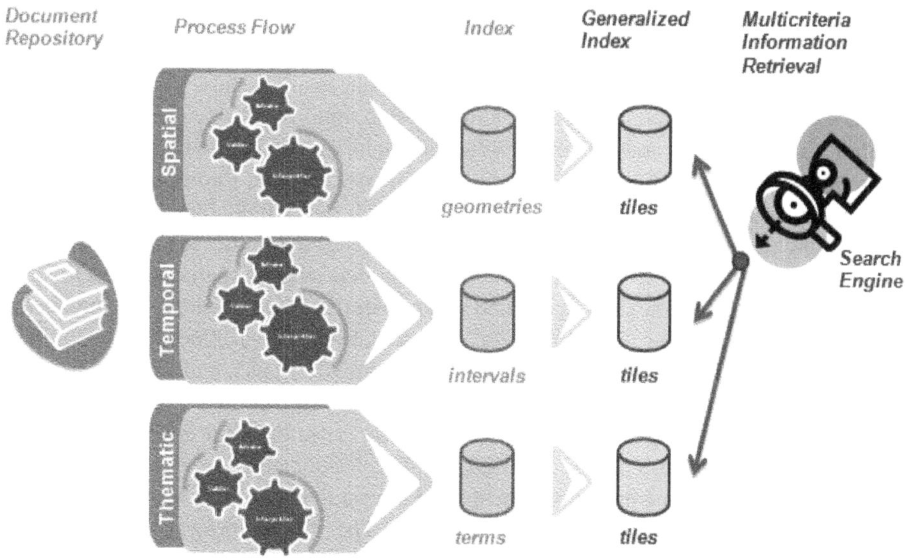

Figure 1.5. *Spatial and temporal generalization for multi-dimensional and multicriteria IR*

We now have models of representation and processing of normalized data for every geographic dimension. We design and implement a multicriteria IR meta-engine combining the results of the spatial, temporal and thematic IRSs (Figure 1.5). These are the generalized spatial and temporal IRSs. The thematic IRS is limited for now to a "basic" engine corresponding to the Terrier IRS [OUN 05].

Such a tool of multicriteria IR targets at least two categories of users. The experts benefit from new operators whose objective is to associate a stronger power of expression with each criterion of the query and thus influence the results of aggregation algorithm. Occasional users will employ predefined aggregation functions that do not need particular skills.

1.4.3. *Applications*

This study is, at the same time, motivated by and contributing to numerous projects (Figure 1.6). The GEOSEM2 project (*Interdisciplinary Program Information Society – CNRS*) has allowed us to experiment with the TAL platform LINGUASTREAM [BIL 06b] and develop spatial information annotating tools in the specific context of document repositories digitized by the Pau Pyrénées Multimedia Library (MIDR). In the continuity of these first results, the PIV project (*Pau Pyrénées agglomeration Community Project*) consists mainly of the development of two process flows dedicated, respectively, to the extraction of spatial and temporal information contained in textual documents. In the frame of the GEOTOPIA project (*CNRS TGE–ADONIS Program – Innovative tools of digital processing for the valuation and diffusion of data*), we adapt and encapsulate these processes in Web services dedicated to spatial and temporal-annotations of documents in a Web platform of sharing and collaborative enrichment of archival data. The GEONTO project (*ANR-07-MDCO-005-01 Project – Flow of Data and Knowledge*) leads us to use the "public" document repository digitized by MIDR for the purposes of enriching a geographic ontology. This geographic ontology, resulting from the processing of numerous resources, is then integrated into the spatial process flow for a better typing of the toponyms and consequently a more adapted approach of disambiguation during the search for representations in gazetteer-like geographic resources (dictionary of place names, describing their physical characteristics – nature or geo-localization of places – as well as social and economical). The GEOCIME project (*64 – General Council Project*) targets the development of an educational application allowing children and their educators to study on collections of multimedia documents before, during and after a field trip. It is based on indexing tools resulting from these different projects. The CIDRI project (*Pau Pyrénées agglomeration Community Project*) is the extension of the PIV project. It aims to generalize spatial and temporal indexes in order to develop an IR prototype combining spatial, temporal and thematic criteria (limited to terms). The MOANO project (*ANR-2010-CORD-024-01 Project – Content and Interactions*), in its GIR aspect, targets the construction and integration of a domain-specific ontology dedicated to botany. This ontology serves as a core model in the generalization of indexes of the thematic process flow. The thematic point of view on the document corpus used is here limited to botany. Finally, the IR engine prototypes PIV^1, PIV^2 and PIV^3 result from these different pieces of study. PIV^1 proposes a spatial IR engine and a temporal IR engine based, respectively, on fine spatial footprint and time stamp representations. PIV^2 proposes a spatial IR engine and a temporal IR engine based, respectively, on generalized representations corresponding to spatial and temporal tilings. Thus, PIV^2 implements the vectorial IR model in both prototypes. PIV^3 proposes a multicriteria IR engine using spatial PIV^2, temporal PIV^2 and Terrier IRS [OUN 05]. PIV^3 then aggregates the results coming from these different IRSs.

All these applications are also motivations for our approach. We can thus say, by recalling our research objectives mentioned earlier, that:

– our models support the symbolic and numeric representation of spatial and temporal data, as well as our spatio-temporal processes of extraction, interpretation and indexing are put to the test in the frame of the GEOSEM2, GEOTOPIA, GEONTO and PIV projects;

– our spatial and temporal IR models are put in practice in the frame of the PIV, GEOCIME and PIV^1 projects;

– our models of generalization and representation of indexed data as well as our models of multicriteria IR are put to experiment in the frame of the CIDRI, PIV², MOANO and PIV³ projects.

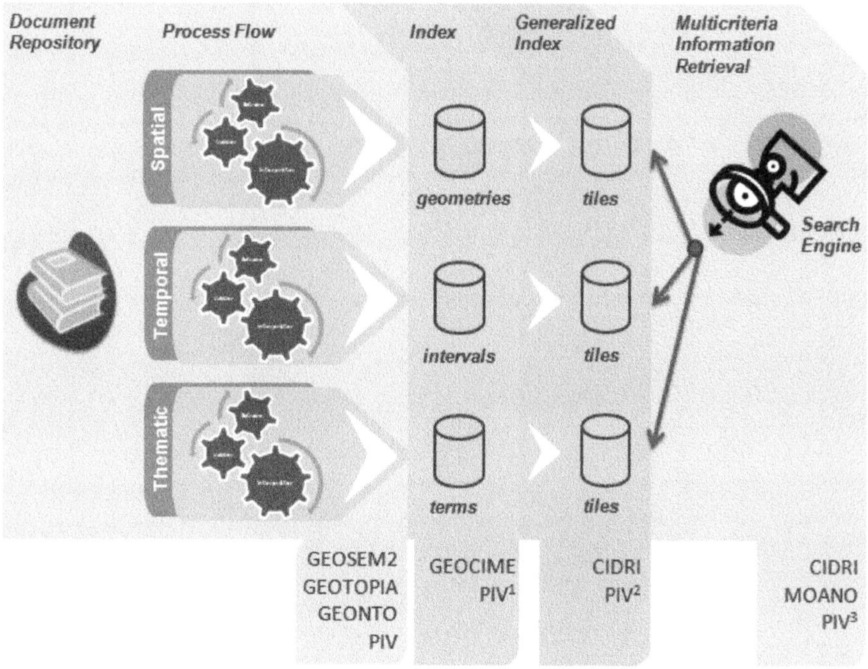

Figure 1.6. *Integration of the study into projects supported by the ANR, CNRS and local authorities*

2

Spatial and Temporal Information Retrieval in Textual Corpora

2.1. Introduction

Information retrieval (IR) systems intended for the wider public do not offer specific processing of spatial or temporal information contained within the corpora or search criteria. Nevertheless, in numerous cases, these pieces of information could play an important role in the calculation of the relevance of a document [TEI 11]. Consideration of the semantics of spatial and temporal expressions could enable a finer processing of expressions such as "*musical instruments in the vicinity of Laruns at the beginning of the 19th Century*". Most of the time in IR, however, documents are processed from the viewpoint of their textual content as mere "bags" of independent words. Moreover, beyond the textual content, document-specific information could be taken into consideration such as the structuring in sections and paragraphs, for example.

Our context relative to textual corpora with "territorial" denotations is specific. On the one hand, spatial and temporal references are frequent and, on the other hand, the document repositories are sufficiently stable and homogeneous to warrant specific *back-office* processing. Our work is thus different from classic IR since it aims a thorough processing of content: specific process flows target the recognition followed by the interpretation of spatial and temporal information. From a structural point of view, the documents at our disposal are acquired from basic digitalization efforts integrating only character and paragraph recognition. They are in text format and are generally composed of several tens or hundreds of pages. This is the reason why we believe the entry point in the corpus cannot be the document itself and we propose working with paragraphs as document units.

As recommended by Clough *et al.* [CLO 06], we deal independently with spatial and temporal dimensions: this way, the single-dimension IR and the operation for adding documents in the corpus remain efficient [MAR 05]. The contributions presented in this chapter are the result of the work undertaken by the T2I team of the LIUPPA laboratory, in the framework of the GEOSEM2, PIV and GEOTOPIA projects (see Chapter 1). This work is mainly discussed in [LEP 07, GAI 08, LES 07].

This chapter is organized into five sections. Following this introduction, section 2.2 recalls the research questions raised in Chapter 1, which will be studied here. Section 2.3 presents propositions found in published literature dealing with the problem of space and time in the processes of annotation and indexing of texts as well as in those of IR. Section 2.4 describes our propositions for indexing together with the spatial and temporal IR. An overview of our contributions and their perspectives are presented in section 2.5.

2.2. Review of challenges, hypotheses and research objectives

Here, we review a subset of issues, hypotheses and research objectives described in Chapter 1. The main issue dealt with in this chapter is thus expressed in the following question: *What models of representation and retrieval of spatial and temporal information should be proposed for the access by geographic content to textual corpora?*

The retained working hypotheses are the following:

– A dedicated spatial information retrieval system (IRS) gives better results than a classic thematic IRS for IR composed of only spatial criteria.

– A dedicated temporal IRS gives better results than a classic thematic IRS for IR composed of only temporal criteria.

– A "rough" coupling of spatial, temporal and thematic IRSs gives better results than a classic thematic IRS for multicriteria IR despite the numerous possible linked biases.

Finally, the research objectives corresponding to the presented issues are mainly aimed at the design and implementation of:

– models supporting the symbolic and numeric representation of spatial and temporal data;

– process flows of spatial and temporal information extraction, interpretation and indexing;

– spatial and temporal IR models.

The contribution concerns first of all the proposition of symbolic representation models of spatial and temporal data. It also concerns the processes aiming at the extraction and semantic analysis of such data in order to build the corresponding annotations [LEP 07, GAI 08]. As an extension to this approach, the following propositions target the enrichment of these symbolic annotations by numeric representations (geometries and calendar periods), generally approximated, stored in the indexes as entry points to noun phrases and corresponding document fragments. They also cover the matching of document fragments with the user need expressed in such an IR context [SAL 07a, LEP 07, SAL 08, SAL 09].

2.3. Spatial and temporal information in textual documents: literature review

We propose a summarized view of the work relative to modeling and reasoning followed by linguistic processes in a context of spatial and temporal information annotation, indexing and retrieval in textual documents.

2.3.1. *Geographic information in text and IR*

Several authors propose a *molecular* definition of geographic information [USE 96, GAI 01, LON 05, LOU 08a]. According to these authors, geographic information links a space, often a time and sometimes descriptive properties. They use a metaphor from chemistry by underlining the atomic character of the spatial, temporal and descriptive components of geographic information (see Figure 1.2). We work with geographic information expressed in natural language. This information is spread out among the lines, which makes its recognition and interpretation difficult for the necessary numeric representation in the IR phase. The geographic information must therefore be identified and converted into data allowing us, to take advantage of their specificity. A process based on the recognition of spatial, then temporal, named entities (NEs) followed by semantic analysis of the text allows the detection of some of the spatial (or temporal) information of a document and its association with a symbolic representation: for example, "to the south of Pau" is specified by a relation of orientation applied to the municipality of Pau and "beginning of January 2010" is represented by a relation of inclusion applied to the month of January 2010. Nevertheless, in order to support the operations of IR, it is then necessary to calculate a numeric representation (geometry and calendar period) corresponding to every piece of information recognized and analyzed in this manner. Moreover, the pieces of detected information can be subjective or dependent on the context in which they are invoked and as a result, the associated numeric representations always imply a certain approximation.

2.3.2. *Named entities*

The notion of NE is widely associated with the establishment of the evaluation campaigns of Message Understanding Conferences (MUC) systems [CHI 97]. These campaigns mainly target the recognition and categorization of textual units referring to the *names of people, businesses and places* as well as other phrases such as *dates, time, monetary units and percentages*, which can be recognized by specific grammar rules [GRI 96, POI 03].

Chinchor [CHI 97] associates the *place* NE category with a place name having a political or geographical connotation (municipality, district, county, region, country as well as toponyms, hydronyms, oronyms, etc.), generally described in external resources. Furthermore, in works of research relative to translation, Bauer [BAU 85] associates proper nouns with six distinct categories, one of which can be linked to historical periods: *praxonyms* are historical facts, diseases or cultural events. Textual documents thus contain many types of temporal information: NEs of type date (more or less complete calendar information: "summer of 1860") and NEs of type praxonym ("Saint-Barthélemy", "Great Irish Famine") also having corresponding calendar periods. We use the calendar term to refer to information determining the beginning, length, and order, and which may be described by years and their divisions.

The recognition of NEs, named entity recognition (NER), consists of processing a stream of words coming from an earlier lexical analysis. A NE detector generally uses a machine learning or an ad hoc rule-base approach to detect and categorize NEs [POI 03]. The learning-based NER is done from texts manually labeled by experts: statistical analysis methods (the texts are considered to be a stream of characters) allow the construction of generic patterns that can be used on a bigger corpus. The ad hoc approach is based on lexical patterns constructed manually with the help of experts: "a proper noun preceded by the preposition *at*, is potentially a place" is an example of such a pattern. These patterns are then applied to a corpus.

There are numerous tools for the automatic recognition of NEs: *GATE ANNIE*[1], *LingPipe*[2], *OpenCalais*[3], *Stanford NER*[4] and *OpenNLP*[5] are qualified as being generalist and aim at the annotation of several categories of NEs, *MetaCarta*[6] and *Yahoo!Placemarker*[7] target NEs corresponding to places while *GuTime*[8] and *HeidelTime*[9] are dedicated to the recognition of NEs corresponding to dates. Concerning *Brat*[10], it is a graphical environment of manual annotation and edition of NEs.

---

1 http://gate.ac.uk/ie/annie.html/.

2 http://alias-i.com/lingpipe/.

3 http://www.opencalais.com/.

4 http://www-nlp.stanford.edu/software/CRF-NER.shtml.

5 http://opennlp.apache.org/.

6 http://www.metacarta.com/.

7 http://developer.yahoo.com/geo/placemaker/; http://developer.yahoo.com/boss/geo/.

8 http://timeml.org/site/tarsqi/modules/gutime/.

9 http://dbs.ifi.uni-heidelberg.de/index.php?id=106/.

10 http://brat.nlplab.org/.

Regarding spatial NEs, we can quote the CasEN project [MAU 11] that defines different typologies of toponyms used in finite state transducers for NER. Other pieces of work [BUS 08, BOU 09] proceed to categorize toponyms after their identification. The approach proposed by Bouamor [BOU 09] mainly takes advantage of the structure of documents: for example, in the collaborative Wikipedia encyclopedia, the identification of NEs is done in the *title* and their categorization is based on the analysis of the first sentence of the corresponding *description* or *category* sections. The approach proposed by Buscaldi and Rosso [BUS 08], and Buscaldi [BUS 09a] aims, in particular, at the disambiguation of recognized toponyms.

Martins *et al.* [MAR 08b] and Alonso *et al.* [ALO 11] describe the difficulties as well as the interests related to the recognition of spatial and temporal NEs in the frame of IR.

We have considered a hybrid approach that, like Buscaldi and Rosso [BUS 08]as well as Bouamor [BOU 09], marks not only place names but also, as proposed by Maurel *et al.* [MAU 11], finds, through the use of external resources (here ontological), associated terms in order to analyze their geographic scope (e.g. populated place, road, stream, park). We present the modeling languages of such information in the following section.

2.3.3. *Modeling languages*

Most of the modeling languages of geographic information are based on Extensible Markup Language (XML) that allows the structuring of information. Let us take the example of the sentence "*Henry Russell admired the vineyard of Jurançon in the late summer of 1856*" with the markup of NEs according to the ENAMEX standard[11] (proposed within the *MUC* campaigns in the 1990s), illustrated in Listing 2.1.

```
<ENAMEX TYPE="PERSON">Henry Russel </ENAMEX>
admired the vineyard of
<ENAMEX TYPE="LOCATION">Jurançon </ENAMEX>
in the late summer of
<TIMEX TYPE="DATE">1856</TIMEX>
```

Listing 2.1. *Example of ENAMEX markup ("Henry Russell admired the vineyard of Jurançon in the late summer of 1856")*

2.3.3.1. *Spatial information markup languages*

There exist different modeling languages of spatial information, each with a particular objective: the exchange of data in the case of Geography Markup

11 http://cs.nyu.edu//cs/faculty/grishman/NEtask20.book_6.html/.

Language (GML), the visualization of data with Keyhole Markup Language (KML), or annotation in the case of SpatialML and ISO-Space.

A widespread spatial modeling language of type interchange format is GML. Defined by the OGC[12], allowing the storage of geographic objects, in particular their corresponding geometries (numeric representations). In addition, GML has been designed to be used within other derived languages of XML. It only handles numeric representations (and therefore not symbolic representations). In the example of "to the south of Pau", a Polygon-type representation (<gml:Polygon>) is associated with it together with the latitudes/longitudes corresponding to the different points.

KML[13] is another widely known spatial modeling language of interchange format type; also defined by the OGC, it is mainly used in *GoogleMaps* and *GoogleEarth*. Like GML, it handles numeric representations, however, it is capable of describing styles associated with the visualization of these representations. Thus, in the example of "to the south of Pau" it associates a representation of type Polygon (<Polygon>) together with the latitudes/longitudes of the different corresponding points. Moreover, information relative to the visualization (e.g. color and line thickness) is associated with these coordinates.

SpatialML[14] [MAN 08a] is a spatial markup language supporting knowledge description. It was developed by the American organization MITRE[15]. SpatialML handles the numeric representations of places (the PLACE tag). It also models spatial relations (SIGNAL and LINK tags), which it only stores as symbolic representations (see Figure 2.2 for the different types of relations). Listing 2.2 illustrates the SpatialML code of the representation of "to the south of Pau". As we can see, the city of Pau is identified and associated with coordinates; the relation of orientation (south direction) is also described but there is no associated numeric representation.

```
<SIGNAL id="1" type="DIRECTION">south</SIGNAL>
<PLACE id="2" country="FR" form="NAM" latlong="43.301667N
    -0.368611W">Pau</PLACE>
<PLACE id="3" />
<RLINK id="4" distance="2" direction="S" source="2" target="3"
    signals="1"/>
```

Listing 2.2. *Example of SpatialML markup for (to the south of Pau)*

12 The *Open Geospatial Consortium* (OGC) is an international consensus that proposes open formats for geographic information – http://www.opengeospatial.org/.

13 http://www.opengeospatial.org/standards/kml/.

14 http://www.mitre.org/work/tech_papers/tech_papers_09/09_3827/.

15 http://www.mitre.org/.

Finally, ISO-Space[16] is a spatio-temporal information markup language in texts. This language mainly integrates tags used for describing movement: the <MOTION> tag allows us to describe a change of places, for example.

2.3.3.2. *Temporal information markup languages*

Concerning temporal information, the most widely employed textual modeling language is TIMEX3. It allows the association of numeric representations with temporal information in the standardized ISO-8601 [MAN 03] format with, for example, tags for beginning and end of duration.

There is also an equivalent to SpatialML: TimeML[17] [PUS 03, PUS 05]. TimeML uses TIMEX3 for the markup of temporal data. It tags temporal relations but only associates them with symbolic representations. Listing 2.3 illustrates the TimeML code of the representation of "beginning of January 2010". As we can see, the date "January 2010" has been identified and a numeric representation has been associated with it; the temporal relation of inclusion (beginning relation type) has also been identified and described, however it does not have an associated numeric representation.

```
<SIGNAL sid="s1">beginning</SIGNAL>
of
<TIMEX3 tid="t1" type="DATE" value="2010-01">
January 2010
</TIMEX3>
<TLINK eventInstanceID="ei1" relatedToTime="t1" signalID="s1"
    reltype="BEGINS"/>
```

Listing 2.3. *Example of TimeML markup of (beginning of January 2010)*

ISO-TimeML[18] [PUS 10], first designed for English and then adapted to other languages such as Italian, Chinese and Korean, is a recent proposal that integrates TimeML for a standardized annotation of time and events.

The SpatialML and TimeML languages were developed to annotate, in other words tag and describe, spatial and temporal information in texts. Numeric representations are associated with places and dates whereas symbolic representations describe the detected spatial and temporal relations. In the following section, we briefly present the main pieces of work relative to reasoning on spatial and temporal relations.

16 http://sites.google.com/site/wikiisospace/.

17 http://www.timeml.org/site/index.html.

18 http://en.wikipedia.org/wiki/ISO-TimeML.

2.3.4. *Reasoning*

We are interested in the work relative to qualitative reasoning. This concerns cognition, in other words the way to make knowledge explicit, from a common sense point of view: for example, without using Euclidean geometry for the qualitative reasoning relative to the spatial dimension. Its origin lies in the works of Allen [ALL 84, ALL 91] on qualitative temporal reasoning. Allen proposes a first classification of the relations between intervals of time [ALL 84, MUL 04]: Figure 2.1 shows these different temporal relations.

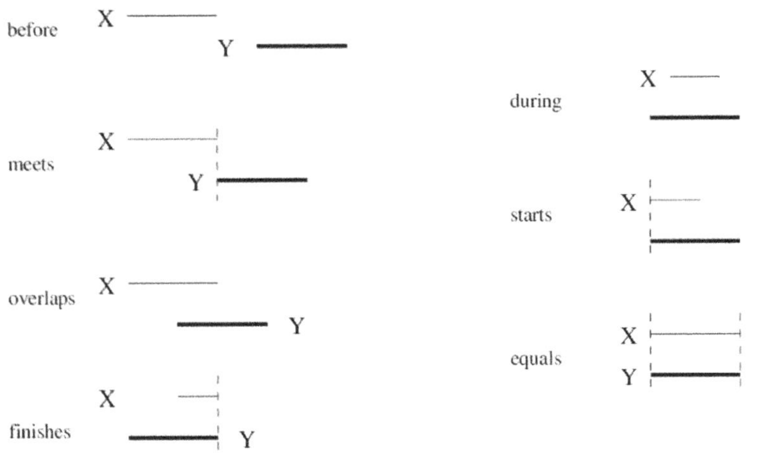

Figure 2.1. *Allen relations [ALL 84] – illustration taken from [MUL 04]*

Proposals for qualitative spatial reasoning have adapted these pieces of work by taking into account the specificities and the greater complexity of spatial information. The *region connection calculus 8* (RCC-8) [RAN 92] is the reference model: it defines eight relations between x and y regions, such as partial overlapping or equality (Figure 2.2 describes the different topological relations). There exist numerous extensions to these pieces of work: for example, Egenhofer *et al.* [EGE 94] propose taking into account linear representations and Szmurlo and Gaio [SZM 98] propose a method for the construction of conceptual neighborhood graphs.

Cohn [COH 96] and Cohn and Hazarika [COH 01] offer a survey relative to qualitative spatial reasoning and categorize, in particular, spatial relations. More recently, Challita [CHA 05a] summarizes the general approach relative to spatial reasoning in the two following steps:

– set the basic spatial entities;

– then, define the atomic qualitative relations that are susceptible to be applied to these spatial entities. These relations are mainly of directional or topological order. Other relations can appear, such as those dealing with the size of objects, the distance separating them or simply their shape.

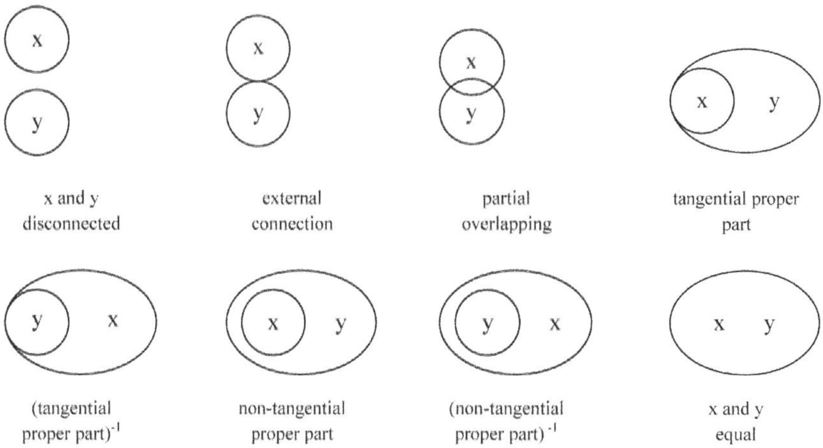

Figure 2.2. *The eight topological relations existing between two x and y regions according to the RCC-8 [RAN 92] model – illustration taken from [LES 07]*

The pieces of work described in [FRE 92], [DEN 97], [LIG 98], [TAL 00] and [LEB 07] are also referential in the field of spatial reasoning.

Parallel to these pieces of work, from a linguistic point of view, Vandeloise [VAN 86] and Borillo [BOR 98] show the particular way humans describe spatial information when it comes to writing: referring to a place involves several elements and these elements' positions within the sentence comply with some rules. They propose the concept of *target/site*: in written language, the target corresponds to the subject of the description and the site constitutes the element of reference. For instance, in the sentence "The car is close to the tree", the *car* is the target and the *tree* is the site. This concept is often used in travelogues. The target is defined here by one or more sites associated with topological, adjacency, orientation or distance indicators or a combination of these. For instance, the text "Beyond Pau, the landscape becomes dull, [...] I have walked from Assat to Eaux-Chaudes...", taken from the book *Voyages aux Pyrénées – Pimientos editions*, describes two places: (1) *the surroundings of Pau* whose place of reference is Pau and (2) *the zone between Assat and Eaux-Chaudes* whose two places of reference are Assat and Eaux-Chaudes. The indicators used here are the indicators of adjacency.

To annotate the information as we have seen in the previous sections, it first has to be identified in the texts and interpreted during the extraction phase. This phase allows a markup of the text in an iterative way in order to constitute indexes later on.

2.3.5. *Linguistic processing*

The processes of information extraction target all the information within a textual document (*full-text* indexing) or only specific elements of information (*targeted* indexing). In the first case, the extracted terms are weighted via so-called statistic approaches: all the terms within a document are processed [MAN 08b, Chapters 2–4]. In the second case, to the contrary, the extraction of the information is based on pre-defined linguistic rules in order to target only particular pieces of information that are generally unweighted [GAI 03, ABO 03].

A spatial and temporal textual content process flow is generally composed of the three following modules:

– *Recognition of NEs*. Lexical analysis allows the conversion of a stream of characters into a stream of words or terms [BAE 99]: *tokenization* splits the document into words following a predefined list of delimiters. These words are then transformed into lexemes during a lexical and morphological analysis. Some less discriminatory and common words ("at", "to", for example) can be eliminated through a so-called list of stop words (or *stoplist*). Finally, an NE detector tags the candidate entities.

– *Validation of NEs*. Knowledge repositories are used for the validation of the candidate NEs.

– *Interpretation of NEs*. Syntactical analysis, based on the rules of grammar, detects relations between lexemes. Finally, a semantic analysis applied to such groups of lexemes ("south of Pau", "torrent of Pau", "beginning of January 2010", for example) allows the processing of adapted interpretation rules. Knowledge repositories are used to disambiguate and then associate representations with the entities.

NLP platforms such as GATE [CUN 02], LINGUASTREAM [BIL 06b], MIRACLE [LIU 06], SXPIPE [SAG 08] and UIMA [FER 04] are dedicated to linguistic processing, and are therefore well adapted to these particular processes. A comparative study is presented in [SAG 08].

Let us note that we address these processes relative to the extraction of information contained within textual documents in the particular framework of IR. Spatial and temporal indexes must therefore be established, which will serve as a support for the IR.

2.3.6. *GIR: systems and similarity measure models*

The main projects relative to GIR are presented hereafter in chronological order of publication:

1) GIPSY [WOO 94], for *Georeferenced Information Processing System*, proposes a method of indexing textual documents. This method is based on the aggregation of geometries corresponding to spatial entities. This aggregation is used to identify the representative geographic zones of the document in view of its indexing.

2) GEOSEM [BIL 03], for *Geographic Semantic*, aims at the semantic processing of geographic information contained in documents such as texts, maps, tables or graphs.

3) SPIRIT [VAI 05], for *Spatially Aware Information Retrieval on the Internet*, allows the retrieval of Web pages describing places or geographic zones specified in a query.

4) GRID [VAL 06], for *Geospatial Retrieval of Indexed Documents*, supports IR in texts. It combines keywords and zones of interest in a cartographic search interface.

5) DIGMAP [MAR 07], for *Discovering our Past World with Digitized Maps*, aims at the promotion of documents relative to cultural and scientific heritage, such as digital libraries of old maps.

6) GEOTRACKER [CHE 07], for *Geospatial and Temporal RSS Tracking*, supports intelligent aggregation and dissemination of RSS feeds.

7) STEWARD [LIE 07], for *Spatio-Textual Extraction on the Web Aiding Retrieval of Documents*, uses spatial information for the retrieval and visualization of information contained within unstructured textual documents.

8) PIIR [LOU 08b], for *Prototype pour l'Interprétation d'Itinéraire dans des Récits de voyages*, is a system of automatic interpretation of itineraries in travelogues.

9) PIV [GAI 08], for *Pyrénées Itinéraires Virtuels*, aims at the promotion of heritage document repositories: newspapers, novels, 19th Century travelogues, processed on the basis of spatial, temporal and thematic criteria.

10) WatWasWaar [LIB 08], for *What has taken place here?* or *What was true?* This title with a double meaning illustrates the fact that history can be *colorful*. Focusing on geographic and historic information (maps and other heritage documents relative to Dutch territory), WatWasWaar processes spatial, temporal and thematic criteria.

11) CITER [PFO 09], for *Creation of a European History Textbook Repository*, supports IR in the textual content of history books.

12) GEOOREKA [BUS 09b], for *Geographically Enhanced Web Search Engine*, targets newspaper extracts relating to a particular geographic zone.

13) SINAI [GAR 09], for *Systemas intelligentes de acceso a la información*, supports IR in textual documents.

14) Document Trajectory [STR 10], for *Document Trajectory Extraction*, explores textual Web documents (collections of Wikipedia articles, for example) and presents event/place couples in a chronological order.

15) Local Search [BRI 10] is a GIR system, processing textual documents, which for the spatial dimension uses an ontology-based index.

16) Mondeca [TEI 11] is a GIR system handling textual documents and offering the possibility of combining thematic and temporal search criteria.

System	Reference	Extraction	Index		
			Space	Time	Theme
GIPSY	[WOO 94]	–	+	–	–
GEOSEM	[BIL 03]	LinguaStream	+	+	+
SPIRIT	[VAI 05]	Gate	+	–	+
GRID	[VAL 06]	Gate	+	–	+
DIGMAP	[MAR 07]	–	+	+	+
GEOTRACKER	[CHE 07]	Miracle	+	+	+
STEWARD	[LIE 07]	–	+	–	+
PIIR	[LOU 08b]	LinguaStream	+	–	–
PIV	[GAI 08]	LinguaStream	+	+	+
WatWasWaar	[LIB 08]	–	+	+	+
CITER	[PFO 09]	Gate	+	+	+
GEOOREKA	[BUS 09b]	–	+	–	+
SINAI	[GAR 09]	Gate	+	–	+
Document Trajectory	[STR 10]	UIMA	+	+	–
Local Search	[BRI 10]	LingPipe	+	–	+
Mondeca	[TEI 11]	–	–	+	+

Table 2.1. *GIR projects: information extraction platforms implemented and geographic dimensions indexed – completed extract of [PAL 10c]*

As shown in Table 2.1, these systems of extraction, indexing and retrieval of geographic information deal with the spatial dimension in priority. Concerning the extraction of the information (see Table 2.1, Extraction column), the most used platforms are GATE, LINGUASTREAM and UIMA. Only the Local Search system integrates the NER tool LingPipe [STO 08]. Let us note that only the GEOSEM and PIV systems deal with spatial and temporal relations during the extraction phase, while Local Search only processes spatial relations of inclusion. The visualization of the indexed documents is an issue addressed by the majority of prototypes. Even though it is not at the heart of this contribution, let us quote the GapVis[19] interface as an example, which proposes an interactive reading of textual documents via synchronized map, text and timeline displayers.

The models of spatial similarity measure implemented in these systems are primarily based on the notion of overlaying between the geographic zone relative to the query (query footprint) and the zone relative to the evaluated document fragment (document fragment/unit footprint). These measures correspond to a form of qualitative reasoning but are based on the topological model of RCC-8, fundamentally qualitative, making the approach hybrid. They are used as relevance ranking tools in GIR. Figure 2.3 illustrates elements of spatial similarity measure whose most used approaches are also presented in [AND 10]:

– Euclidean distance: formula [2.1] shows that the shorter the distance (d) between the query (Q) and query/document overlap (O) footprints, the more relevant a document is to the query location (see Figure 2.3). Jones *et al.* [JON 01] and Vaid *et al.* [VAI 05] present similar approaches to measure the proximity between the spatial footprint expressed in the search criteria and those recognized in candidate documents.

$$Similarity(D_f, Q) = \frac{1}{d} \qquad [2.1]$$

– Overlapping surface: formula [2.2] shows that the greater the overlapping surface (O) between the query Q and document D_f spatial footprints, the higher the relevance of the document to the query (see Figure 2.3). Jones *et al.* [JON 01] and Larson and Frontiera [LAR 04] present similar approaches to measure the overlapping surface between the geographic zone expressed in the search criteria and those recognized in candidate documents.

$$Similarity(D_f, Q) = O \qquad [2.2]$$

– Overlapping ratio: formula [2.3] shows that the more the value of the overlapping surface ($O * 2$) is similar to that of the query Q and document D_f footprint surfaces together, the more the document is relevant to the query location (see Figure 2.3). Hill [HIL 90], Walker *et al.* [WAL 92], Jones *et al.* [JON 01] and Larson and Frontiera [LAR 04] present similar approaches to measure the overlapping ratio. The works of Beard and Sharma [BEA 97], reflected in [LAR 04], take into account the topological relations existing between the query Q and document D_f spatial footprints: formula [2.4] is applied when Q contains D_f; formula [2.5] is applied when D_f overlaps Q ($(O \cap Q)\%$ equals the percentage of overlap of O with Q and $\neg(O \cap Q)\%$ equals the percentage non-overlap of O with Q) so that the percentage of

19 http://gap.alexandriaarchive.org/gapvis/.

non-overlap region lower the similarity; formula [2.6] is applied when Q is contained within D_f.

$$Similarity(D_f, Q) = 2 \cdot \frac{O}{Q + D_f} \quad [2.3]$$

$$Similarity(D_f, Q) = \frac{D_f}{Q} \quad [2.4]$$

$$Similarity(D_f, Q) = \frac{(O \cap Q)\%}{\neg(O \cap Q)\% + 100} \quad [2.5]$$

$$Similarity(D_f, Q) = \frac{Q}{D_f} \quad [2.6]$$

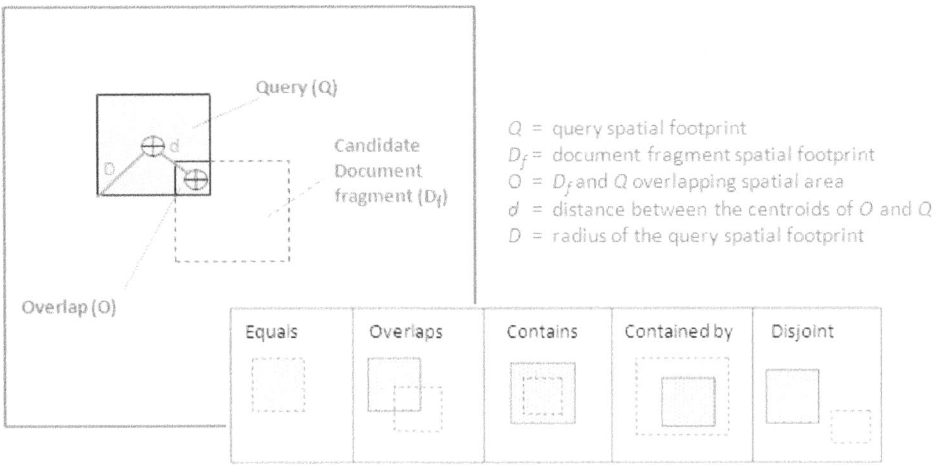

Figure 2.3. *Elements of measure of spatial similarity*

The work relative to the measure of temporal similarity is less diverse. It is mainly based on the notions of inclusion and overlapping between timestamps, as described in [ALL 84], and usually provides Boolean results for those involved in the systems listed in Table 2.1.

However, Teissèdre *et al.* [TEI 11] propose a temporal relevancy calculation heuristic with the double objective of (1) ranking according to relevance the results of a retrieval and (2) integrating the documents that are non-relevant in the sense of Allen but sufficiently close to the search period to possibly interest the author of the query. Thus, the criteria used in the calculation of temporal relevance as proposed in [TEI 11] integrate:

1) the overlapping surface;

2) the measure of the distance between calendar references appearing, respectively, in the query and in the documents (in other words the timeline separating two calendar periods whose overlapping is nil);

3) the measure of the proportion ratio from the view point of the range of the entities;

4) the measure of the distance weighted by the range of the entities;

5) the measure of the overlapping ratio.

Alonso [ALO 08] proposes a formula that, after standardization of the range of the temporal entities and of the number of calendar expressions, calculates the number of standardized expressions present at the same time in the query and in the document.

The majority of GIR projects listed in Table 2.1 are based on geographic information systems (GISs) for the storage and parsing of indexes followed by the calculations of spatial similarity. Similarly, they are generally based on database management system (DBMSs) for the storage and parsing of indexes followed by the calculations of temporal similarity. The approximation of spatial and temporal entities with numeric representations is necessary for the implementation of such calculations.

2.3.7. *Evaluation campaigns, corpora and resources*

An early evaluation of the different IR approaches was deemed necessary. Several criteria have been defined by Cleverdon in [CLE 63]:

– the ease of use of the system;

– the access/storage cost;

– the presentation of results;

– the capacity of the system to select relevant documents.

For this last criterion, Cleverdon has proposed two now-recognized measures used in most evaluations of IR methods. These are recall R and precision P:

$$R = \frac{\textit{Number of relevant documents selected}}{\textit{Number of relevant documents}} = \frac{RS}{R}$$

$$P = \frac{\textit{Number of relevant documents selected}}{\textit{Number of documents selected}} = \frac{RS}{S}$$

Recall evaluates the capacity of a system to select all the relevant documents in the collection (are all the relevant documents selected?) while *precision* evaluates the capacity of the system to select only relevant documents (are all the selected documents relevant?). Other measures based on these two have been proposed [SAN 10]. For instance, mean average precision (MAP) corresponds to the average precision calculated on a given set of test queries.

The field of IR is characterized by a long history of evaluation [VOO 02]. A way to evaluate the IRSs is based on the definition of a "campaign" that occurs in the following way:

1) The organizers spread a call for participation, which presents the proposed IR tasks. For example, an ad hoc task requires to retrieve a list of relevant documents for a given query. In contrast, for a *question answering* task, the aim is to retrieve a piece of information answering to a given query. For the query "beaches of Anglet", we would obtain a list of documents dealing with this subject for the ad hoc task, whereas we would get the list of the beach names of Anglet for the *question answering* task.

2) The interested IRS designers register to the tasks of their choice. They are then referred to as *participants*.

3) The organizers provide a *corpus* of documents and 25+ *topics* representing information needs (i.e. detailed queries with description and narrative).

4) The participants process the corpus, submit the topics to their IRS and then pass the obtained results, also known as *runs*, to the organizers (e.g. per topic document list ranked by decreasing relevance).

5) The organizers constitute a set of relevant documents for each topic: the *relevance judgments*. They then check participants' results against these relevance judgments by means of predefined appropriate measures. The computed value represents the effectiveness (i.e. measurement of result quality) of the IRS for the considered topic. Aggregating all the scores obtained by the IRS for each of the 25+ topics (e.g. averaging over them) leads to an overall evaluation score for the IRS.

6) The organizers publish the results of the participants and generally make available the *test collection* (i.e. corpus, topics and relevance judgments). This collection can then be reused later in order to evaluate an IRS outside the campaign framework.

As shown in Figure 2.4, TREC [VOO 05] is a reference campaign in IR allowing us to evaluate IRSs with respect to the thematic dimension. SEMEVAL [AGI 07] and SEMSEARCH [HAL 10] are, in particular, involved in the semantic analysis of textual contents. There is not a lot of published work relative to the evaluation of the two other dimensions of geographic information. The spatial and temporal dimensions have been the object, respectively, of the evaluation framework CLEF ([PET 01], task

GEOCLEF [GEY 05]) and of the evaluation framework TEMPEVAL [VER 09]. For example, GEOCLEF 2008 contains articles from different newspapers (169,477 articles in English, 294,809 in German and 210,734 in Portuguese) and 25 *topics* (queries integrating spatial and thematic criteria).

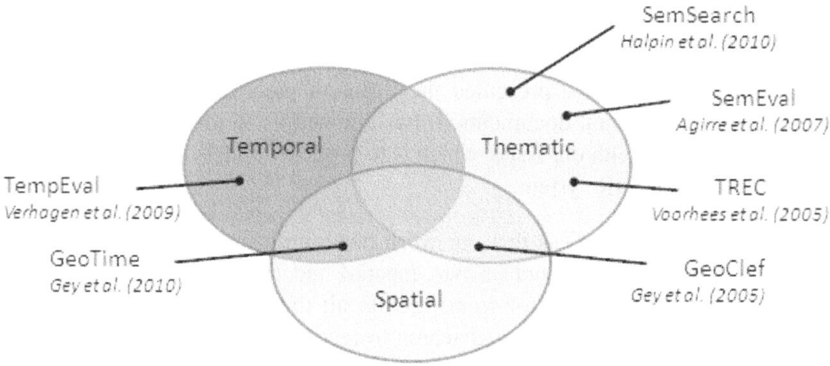

Figure 2.4. *IRS evaluation campaigns*

Other manually annotated reference corpora have been proposed for the evaluation of toponym resolution tasks. Table 2.2 illustrates a few examples taken from [AND 10]. Let us note that these corpora are mainly composed of press articles and that the reference language is English.

	TR-CLEF	TR-RNW	TR-CoNLL	TR-MUC4
Reference	[AND 10]	[AND 10]	[LEI 07]	[LEI 07]
Size of corpus (in tokens)	360,559	6,010	204,566	30,051
Number of documents	321	556	946	100
Toponym instances	5,783	2,338	6,980	278
Distinct toponyms	802	432	1,299	135
Ambiguous distinct toponyms	690	332	–	–
Non-ambiguous distinct toponyms	112	102	–	–
Human annotators	2	1	4	2

Table 2.2. *Corpora dedicated to the evaluation of toponym resolution tasks (table created from data presented in [AND 10])*

Resources or spatial gazetteers are associated with these evaluation campaigns. These resources are mainly composed of lists of toponyms and of corresponding geometries. Numerous complementary pieces of information, such as the type (feature class) of the toponym, can be associated with the description. These resources are necessary for the recognition, validation and interpretation of spatial

NEs in a GIR system. We can mention Alexandria Digital Library (ADL)[20], GeoNames[21], Getty Thesaurus of Geographic Names (TGN)[22], GEOnet Names Server (GNS)[23], Geographic Names Information System (GNIS)[24], Géoportail IGN[25] (BD TOPO and BD CARTO of the French "Institut Géographique National").

2.3.8. *Summary*

In section 2.3, we have presented the different pieces of work relative to the content analysis of textual documents. It has allowed us to identify the elements of the literature dealing with our issue, which is to propose models and methods to build a spatial and temporal IR system.

A result from this study is that our work lies within the field of specialized IR. Indeed, we wish to process only a part (spatial and temporal) of the information contained within texts instead of working with all the terms present in the corpus. Thus, we have to perform a *targeted* search to recognize the significant elements of the text. It is necessary to carry out an interpretation of such elements; indeed, indexing "south of Pau" without knowing the spatial significance of this phrase is not sufficient. The linguistic approach, using the tools of NLP, seems to be the most adapted method to perform this *active* retrieval. This allows us to extract specific phrases and to interpret the semantics of the relations between the terms of each phrase.

However, the data recognized in such a way have to be organized in indexes. It is therefore important to interpret the spatial and temporal content of the textual documents in such a way as to build adequate structures. This is the reason why the work relative to qualitative reasoning allows us to interpret and structure the information initially extracted using linguistic tools. In particular, the concept of target/site seems to be a good starting point for the interpretation and structuring of spatial and temporal information.

In the following section, we propose an ad hoc model of representation and retrieval of spatial and temporal information, but inspired from the existing model, in order to take into account the spatial and temporal relations present in texts. The same applies to the evaluation of our propositions. The testing protocol is similar to the existing protocol, the experimental corpus is in French and the resources cover the toponymy relative to the Pyrenean region.

20 http://www.alexandria.ucsb.edu/gazetteer/.

21 http://www.geonames.org/.

22 http://www.getty.edu/research/tools/vocabularies/.

23 http://earth-info.nga.mil/gns/html/.

24 http://nhd.usgs.gov/gnis.html/.

25 http://www.geoportail.fr/.

2.4. Proposition for spatial and temporal information indexing and retrieval in textual corpora

We present the spatial and temporal IR platform that we have developed. It is mainly based on the spatial and temporal data representation models as well as the spatial and temporal information indexing and retrieval process flows that we designed. Referred to as PIV, this platform is tested on a corpus composed of travelogues relative to the Pyrenean ranges, written in French.

2.4.1. *Reminder and focus on the notion of space and time in "heritage" corpora*

We illustrate the spatial and temporal expressions typical of our corpus with a few examples. We mainly deal with toponymic and calendar expressions.

From a spatial point of view, we aim at the recognition and interpretation of geographic zones mentioned in the texts. Here are some examples of spatial expressions extracted from our documents:

> Pau; the Cerdagne; the inner city of Lourdes; the woods of Perthus; the pastures of Pourtalet; the ridges of the Canigou range; the footpath of Cadeilhan Trachères; in the forest of Iraty; at the foot of the Bastanet pass; above Aragnouet; to the right of Oloron; more than two hours from Oule; toward the Literola lake; near the Pic de Rouille; away from the road of Perthus; not far from Balledrayt hut; the region of Vallier; the slope of Ariège; close to the Saint-Béat and Luchon cantons; on the outskirts of the lowest passes; to the west; the arid land to the south of the Aragon region.

Some expressions refer to the entries in resources relative to the Pyrenean territory (*IGN*, collaborative resources like *OpenStreetMap* or free resources like *GeoNames*): these are the toponymic names (Pau) or extended toponyms specifying the type of the spatial entity thus described (woods of Perthus). We will call these equally *absolute entities* or *absolute features*. Others correspond to an adaptation of one or more absolute spatial entities (next to the Rouille pass), we will call these *relative spatial entities* or *relative spatial features* (RSFs) [SAL 08]. Finally, the spatial entities can be complete or incomplete. Incomplete entities (on the outskirts of the lowest passes) cannot be located on a map since an elementary analysis, limited to the noun phrase, does not allow their geolocalization.

From a temporal point of view, our objective being to identify the periods discussed in the documents, we are only interested in calendar-like temporal expressions. Here are some examples of temporal expressions that can be found in our documents:

> 11 March 1957; July 1957; 11 March; March; spring of 1787; summer of 1799; winter; 2002; 18th Century; 18h00; end of 1892; from 1781; since 1787; 1997–1998; end of August 1801; from 1950 to 1960; early 1950s; toward the end of spring 1800; April and May 1798; first days of autumn; toward the end of 1794 or the beginning of 1795; 12, 15 and 20 July 1916; 1950,52,57; from 29 to 30 September; before the Christian era; since the revolution; during 18 years.

As a first step, we have decided not to include the expressions referring to hours of the day (less convinient level of detail, in our contexts of use), to durations (anaphoric references difficult to analyze) and to historical events (events needing specific thesauri). In the remaining subset, some expressions explicitly refer to a calendar entry: these are the dates ("11 March 1957"; "spring of 1901"; "18th Century"). We will call these equally *absolute entities* or *absolute features*. Others correspond to an adaptation of one or more absolute temporal entities ("end of 1892"; "toward the end of spring 1800"). We will call these *relative temporal entities* or *relative temporal features* (RTFs). The finest temporal range is the day. The temporal entities can be complete or incomplete. Incomplete entities ("March"; "11 March"; "winter") cannot be located on an absolute time scale since they are not associated with a year in an explicit way in the corresponding noun phrase.

2.4.2. *Core spatial model and core temporal model*

Following the *target/site* hypothesis described in the state of the art [BOR 98], we consider that an important part of spatial or temporal information expressed in a text is composed of at least one NE and a variable number of spatial or temporal indicators, specifying its range. The development of the model has also been guided by the study of corpora. Following this study, we have proposed a unified model to describe the spatial features (SFs) and temporal features (TFs). This model is referred to as *core* since it will be used for all the steps of information indexing and retrieval. Figure 2.5 gives an overview of the model.

SFs as well as TFs can be absolute or relative. A relative feature is described recursively from one or more features. Recursion confers to this model the power of expression necessary for the representation of complex entities. Each entity is associated with one or more representations.

An absolute spatial feature (ASF) corresponds to a geolocalizable spatial NE ("the city of Pau", "Laruns") for which there are numerous classifications [HIL 90]. If the administrative entities (cities, roads, for example) are precisely georeferenceable, the entities corresponding to the physical environment (mountains, lakes, rivers, valleys, for example) have sometimes ill-defined limits or are without consensus (e.g. *fiat* and *bona fide* borders [SMI 01]). An RSF corresponds to a site NE accompanied by spatial indications allowing us to identify the target zone: spatial relations are associated recursively with SFs. For example, "north of Pau" is an RSF that defines a relation of orientation with the ASF "Pau". This recursive construction of an RSF allows us to process complex relations containing successively encapsulated topological, orientation and/or distance indications. For more details on the spatial model, refer to [LES 06].

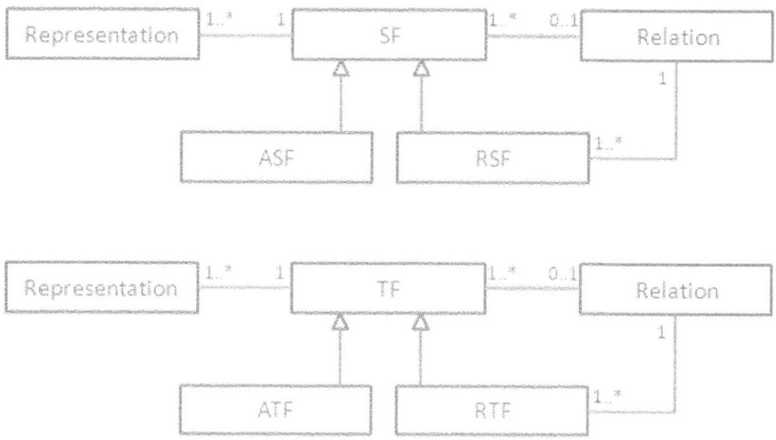

Figure 2.5. *Simplified UML schema of the spatial model and the temporal model, extract taken from [LEP 07]*

An absolute temporal feature (ATF) corresponds to a calendar-NE ("11 June 2000", "summer of 2009"): day, month, season, year, century or era. It can be incomplete (day, month or season, without the year, for example). An RTF corresponds to an NE accompanied by temporal indications allowing us to specify a target period: temporal relations are associated with the TF (absolute or relative) recursively. For example, "around March 2000" is an RTF that defines a relation of adjacency with the ATF "March 2000". For more details on the temporal model, refer to [LEP 07].

We have chosen to represent the SFs and the TFs using geometries and time intervals, respectively. The main interest is to allow, during the retrieval phase, to calculate overlapping rates and, in particular, to use the operators defined in [RAN 92] and [ALL 84], respectively.

2.4.3. *Spatial and temporal relations*

We call *relations* as the indicators applied to SFs or TFs. Our approach is based on the existing models of qualitative spatial reasoning [COH 96, COH 01, MIR 07] but also quantitative, for the calculation of approximated numeric representations [FU 05, LIU 09]. We identify five spatial relations:

– adjacency: unary relation allowing us to describe SFs such as “near SF”, “far from SF” and “the periphery of SF”;

– inclusion: unary relation allowing us to describe SFs such as “the neighborhood of SF”, “the center of SF” and “the peak of SF”;

– orientation: unary relation allowing us to describe SFs such as “south of SF”;

– distance: unary relation allowing us to describe SFs such as “an hour’s walk from SF” and “ 20 km from SF”;

– union and intersection: n-ary relations that link at least two SFs such as “between SF1 and SF2”, “the SF1, SF2, SF3 triangle”, “at the intersection of SF1 and SF2” and “the SF1-SF2 border”.

An RSF is thus defined by one of these relations and at least another SF recursively.

By the same principle, we distinguish five types of temporal relations:

– adjacency: unary relation allowing us to describe TFs such as “about TF”, “around TF” and “the TFs”;

– inclusion: unary relation allowing us to describe TFs such as “in the early TF” and “in the late TF”;

– orientation: unary relation allowing us to describe TFs such as “before TF” and “after TF”;

– interval: binary relation allowing us to describe TFs such as “from TF2 to TF2”;

– enumeration: n-ary relation allowing us to describe TFs such as “9, 10 and TF” (“12 November 1900”, for example).

An RTF is thus defined by one of these relations and at least another TF recursively.

The recursive definition of the SFs and TFs is an advantage of our model not only from a pragmatic point of view but also from a cognitive point of view, since it is close to our mode of reasoning. Two modes of representation are thus determined. A first symbolic annotation recursively transcribes the semantics expressed in the text. A second annotation referred to as numeric corresponds to an approximated representation in the form of a georeference or an interval associated with the SF or the TF, respectively. The numeric representation is related to an IR usage. Other uses of these representations are obviously possible.

2.4.4. *Spatial and temporal indexing process flows: PIV prototype*

In accordance with the specifications of Clough *et al.* [CLO 06], we deal independently with each of the spatial and temporal dimensions and build specialized indexes. This way, the monodimensional retrieval and the management of indexes (addition of new documents in the corpus) remain efficient [MAR 05].

The spatial and temporal process flows are composed of three main processing steps (Figure 2.6). The first step consists of the extraction of the SFs and TFs using a lexico-syntactic process [GAI 08]. This process is supported by the LINGUASTREAM [BIL 03, WID 05] platform. After a classic preliminary segmentation processing of the contents (*tokenization*), it implements a form of active parsing: a candidate *token* tagger identifies those corresponding, respectively, to spatial and temporal information, by using typographical rules as well as lexical resources. A morpho-syntactical analysis gathers tokens to constitute nominal groups corresponding to candidate SFs and TFs ("torrent of Pau", "Tuesday 21 June 2011", for example). These candidate features are composed of a proper noun or a number for the SFs and TFs, respectively. Moreover, they have a more or less high reliability score depending on the existence of an introducer referenced in the lexical resources. Grammatically, these introducers belong to the category of prepositions of place (*in, on, under, next to, near, far*, etc.), prepositions of time (*of, in, since, during*, etc.), adverbs of place (*in the proximity of, around*, etc.), adverbs of time (*around, after, before*, etc.) and adjectives of localization (*back, central, north, higher*, etc.).

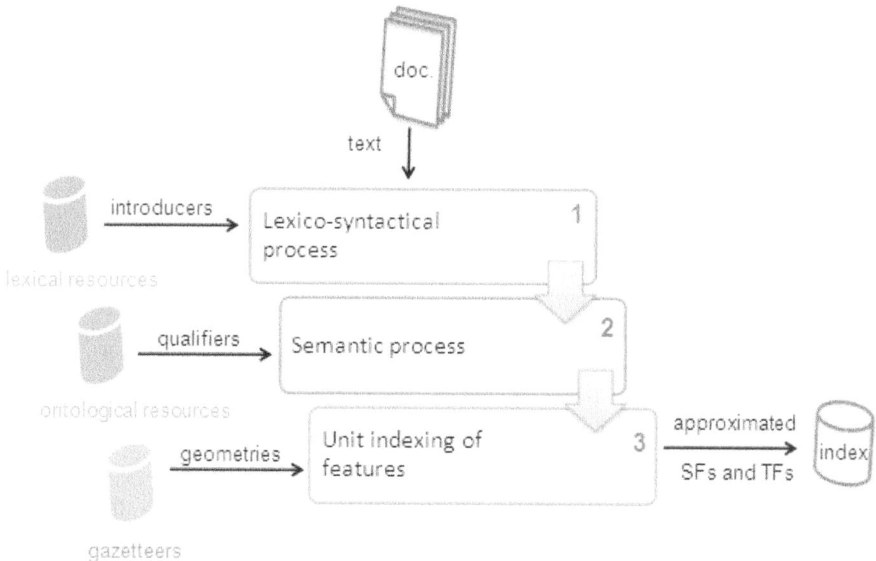

Figure 2.6. *The three main steps of indexing in the PIV prototype*

The second step determines a symbolic representation for each of the candidate SFs and TFs identified earlier. We design semantic patterns and implement them with definite clause grammars (DCGs): a recursive semantic analysis of each SF and TF matches ASFs, RSFs, ATFs and RTFs. Therefore, anytime a spatial or temporal relationship is found out, the corresponding RSF and RTF is analyzed and tagged (i.e., adjacency, inclusion and orientation with an SF or a TF). Similarly, a second set of DCGs is dedicated to the categorization of the SFs and TFs nature (e.g. oronym, hydronym, town and road tags characterize SFs; day, week, month, season, year and decade tags characterize TFs). Thus, for example, the noun phrase "to the north of the torrent of Pau" is annotated as RSF with an *orientation spatial relationship: north* tag and a *category: hydronym* tag. This RSF is defined recursively from the ASF of toponym *torrent of Pau*. Each of these tagging processes is supported by specific ontological resources [KER 09].

The third step interprets the symbolic representations obtained and calculates numeric representations. The interpretation is supported by approximation algorithms that associate timestamps with the TFs [LEP 07] and spatial footprints with the SFs [FU 05, GAI 08, LIU 09]. The spatial approximation process involves resources like the PostGIS[26] GIS, as well as gazetteers (e.g. database BD NYME® of the IGN, GeoNames and local collaborative resources) in order to validate and retrieve each ASF geometric representation. The spatial relationships (e.g. *orientation: north*) are processed and corresponding representations are approximated. GIS operators (e.g. *translation, intersection*) are applied to the reference SF. For the temporal approximation, using the same principle, a period is initially associated with each ATF and then the temporal relations (e.g. *adjacency: toward*) are interpreted and one or more timestamps are calculated from the reference TF. A spatial index and a temporal index are the result of this third step. The spatial index describes each SF with the geometry, the phrase and the identifier of the corresponding paragraph and document. Similarly, the temporal index describes each TF with the period, the phrase and the identifier of the corresponding paragraph and document.

Both spatial and temporal processing flows are described in a detailed way in [LES 07] and [LEP 07], respectively. Each detected SF and TF is represented by an instance of the model shown in Figure 2.5. The calculation of the spatial and temporal representations relative to the RSFs and RTFs is supported by algorithms specific to each modeled spatial and temporal relation. Figures 2.7 and 2.8 schematize the reasoning: the reference site is illustrated in dark-gray and the target is represented in light-gray. It is build by enlargement, reduction and movement of the site.

Figure 2.7 illustrates the different interpretations planned by the implementation for the five spatial relations defined in the core model. For *adjacency*, the bounding box of the reference site, of side-length c, is extended by a factor of $c/2$ on each side. For *inclusion*, the sides of the bounding box of the reference site are reduced by $c/2$.

26 http://postgis.refractions.net.

For *orientation*, two sides are extended by a factor of $c/2$ and the two others by a factor of $2c$. The bounding box is then subjected to a translation of value $1.5c - c/5$ toward the chosen orientation, in order for the target bounding box to keep a small overlapping surface with the original bounding box. The relation of *distance* is similar to that of adjacency. Only the factor is no longer proportional to the bounding box of the reference site but is guided by the distance (x) given in the text. Finally, the relation of *geometric figure* simply describes a new box surrounding the bounding boxes of the two reference sites. These choices of implementation of the recursive function calculating geometric representations are detailed and validated in [LES 07]. We have in particular chosen to prioritize the recall of the system, thus allowing noise.

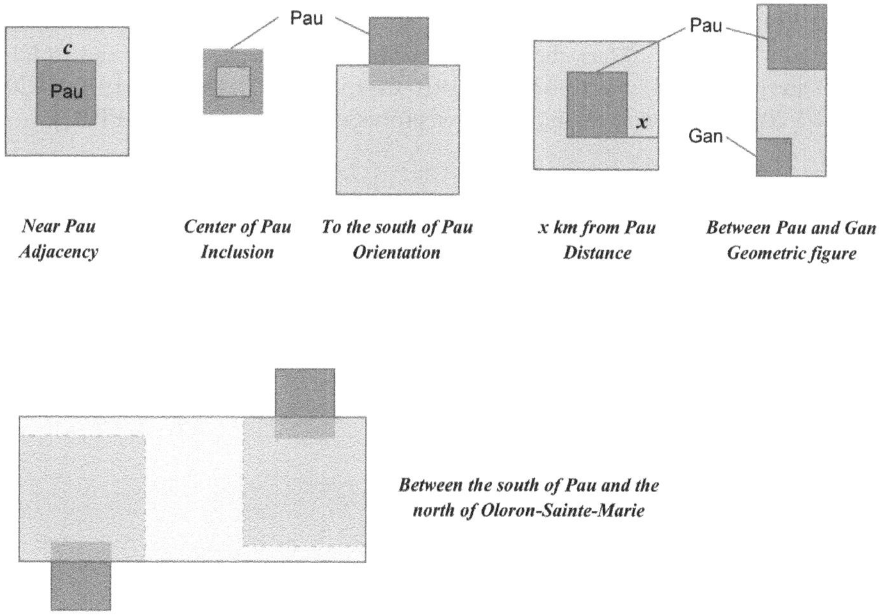

Figure 2.7. *Interpretation of the five spatial relations of the core model – extract taken from [LES 07]*

Figure 2.7 shows an example of recursive interpretation. The sentence "between the south of Pau and the north of Oloron-Sainte-Marie" is interpreted in several steps. A first representation is calculated for "the south of Pau", as well as for "the north of Oloron-Sainte-Marie". Lastly, the final representation is produced from these.

The representations are proposed in the form of bounding boxes. This solution, chosen as the first approximation during the implementation of the PIV prototype, enables fast calculations during the IR phase.

Figure 2.8 illustrates the different interpretations planned by the implementation of the five temporal relations defined in the core model. Concerning *adjacency*, the reference period is extended at the level of its two bounds. This extension depends on the nature of the adjacency ("toward" and "the years") and on the range of the reference period. For *inclusion*, the bounds of the reference period are reduced: the algorithm depends on the nature of the inclusion ("beginning", "middle" and "end") and on the range of the reference period. For example, an inclusion of the first type corresponds to the beginning of the third reference period. For *orientation*, the reference period is extended at the level of one of its two bounds. This extension depends on the nature of the adjacency ("before" and "after") and on the range of the reference period. The relation of *interval* between two reference periods denotes a new period covering the first two: the lower bound of this new period is that of the first period and its upper bound is that of the second period. Finally, the relation of *enumeration* denotes a list of values associated with a reference period such as "12, 15 and 20 July 1916" whose reference ATF site is "20 July 1916" and the associated values are "12 and 15". Thus, the target timestamps representing "12, 15 and 20 July 1916" are [1916-07-12, 1916-07-12], [1916-07-15, 1916-07-15] and [1916-07-20, 1916-07-20]. These recursive interpretation processes are described in [LEP 07].

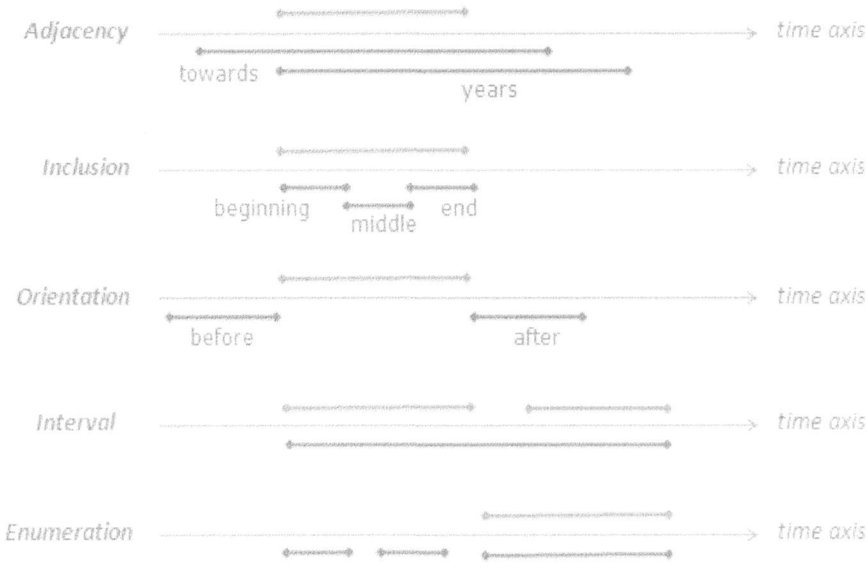

Figure 2.8. *Interpretation of the five temporal relations of the core model – simplified illustration taken from [LEP 07]*

2.4.5. *Spatial and temporal IR: PIV prototype*

This section describes the process of retrieval that uses the indexes created earlier. The process has to facilitate the access to the documents of the corpus in the case of queries having a spatial or temporal dimension. The query can be expressed in free text and, in this case, is subject to the same process as the documents. It is therefore translated into georeferenced areas, respectively, in calendar periods, which will be used for the matching process. Other modes of expression, mainly graphical, are proposed in [LES 07] and [PAL 10a]. The system retrieves document fragments containing spatially or temporally relevant noun phrases with respect to the query.

2.4.5.1. *Spatial IR*

We have tested the models of similarity measure discussed in section 2.3.6 and composed, in an empirical way, a derived ad hoc model adapted to the corpus. This spatial IR process passes through the following steps:

1) *Interpretation of the query.* The query is interpreted using the same process flow as for indexing. SFs are detected and then symbolic and numeric interpretations are calculated.

2) *Calculation of the set of results.* Let Set_{req} be the set of SFs annotated in the query and Set_{doc} the set of SFs annotated in the document. We have $Set_{req} = \{SF_{req}\}$ and $Set_{doc} = \{SF_{doc}\}$. Then, we calculate Set_{res}, which is the set of SFs of Set_{doc} for which the overlapping of their spatial representation and that of an SF of Set_{req} is not empty. We have $Set_{res} = \{SF_{doc}\}$ with $SF_{doc} \in Set_{doc}$ and $\exists\, SF_{req} \in Set_{req}$ such that $representation(SF_{doc}) \cap representation(SF_{req}) \neq \varnothing$. The result of the query contains the set of document fragments to which the SFs of Set_{res} belong.

3) *Calculation of the relevance score of each document fragment in the set of results.* We use the characteristics detailed in Figure 2.3 to measure the similarity between a document fragment D_f and a query Q [SAL 07a].

$$Similarity(D_f, Q) = \frac{Precision(D_f, Q) + Overlapping(D_f, Q)}{2 + Distance(D_f, Q)} \qquad [2.7]$$

Equation [2.7] uses notions of *precision, overlapping* and *distance*. The *precision* score [2.8] evaluates the relevance of a document, in other words if the surface of overlapping O (Figure 2.3: overlapping area, considered to be relevant) occupies a large part of the surface of document D_f. In a similar way, the *overlapping* score [2.8] evaluates the ratio of the surface of query Q occupied by the surface of overlapping area O. The higher this score, the more significant the corresponding document is. Finally, the *distance* score [2.8] evaluates the distance of the centroids of query Q and that of the surface of overlapping area O. The closer these centroids are to each other, the more relevant the document fragment D_f is.

$$\begin{aligned} Precision(D_f, Q) &= \frac{O}{D_f} \\ Overlapping(D_f, Q) &= \frac{O}{Q} \\ Distance(D_f, Q) &= \frac{d}{D} \end{aligned} \qquad [2.8]$$

Equation [2.7] is translated into an SQL query containing GIS functions in order to perform the search and to score the selected document fragments. Finally, an ordered list of document fragments constitutes this result.

If we take the example of a query containing the phrase "to the south of Pau", an RSF is recognized and a georeferenced footprint is computed. This representation is compared to those contained in the index and, for example, a document fragment containing "city of Gan"[27] is retrieved. The computed overlapping area corresponds then to an instance of the core model: *the intersection of the south of Pau and of Gan*. Yet, like this sentence, which, interpreted by a human, can give a *qualitative* relevancy score, its georeferenced representation gives a *quantitative* relevancy score.

We propose a calculation of spatial relevance supported by the bounding box representations, which can be extended to other geometric primitives.

2.4.5.2. *Temporal IR*

The propositions in published literature relative to the calculation of similarity scores in temporal IR are recent (see section 2.3.6). Since 2007 (see [LEP 07]), we have proposed a similar approach to that developed for spatial IR.

Indeed, we have developed a model derived from the spatial IR model discussed above. Therefore the process of temporal IR also involves the following steps:

1) *Interpretation of the query.* The query is interpreted using the same process flow as for indexing. TFs are detected and then symbolic and numeric interpretations are calculated.

2) *Calculation of the set of results.* Let Set_{req} be the set of TFs annotated in the query and Set_{doc} the set of TFs annotated in the document. We have $Set_{req} = \{TF_{req}\}$ and $Set_{doc} = \{TF_{doc}\}$. Then, we calculate Set_{res}, which is the set of TFs of Set_{doc} for which the intersection of their temporal representation and that of a TF of Set_{req} is not empty. We have $Set_{res} = \{TF_{doc}\}$ with $TF_{doc} \in Set_{doc}$ and $\exists\, TF_{req} \in Set_{req}$ such that $representation(TF_{doc}) \cap representation(TF_{req}) \neq \varnothing$. The result of the query contains the set of document fragments to which the TFs of Set_{res} belong.

3) *Calculation of the relevance score of each document fragment in the set of results.* We use the characteristics detailed in Figure 2.9 to measure the similarity between a document fragment D_f and a query Q.

$$Similarity(D_f, Q) = \frac{Precision(D_f, Q) + Overlapping(D_f, Q)}{2 + Distance(D_f, Q)} \qquad [2.9]$$

27 Gan is located 10 km to the south of Pau.

Equation [2.9] uses notions of *precision*, *overlapping* and *distance*. The *precision* score [2.10] evaluates the relevance of a document, in other words if the common period O (Figure 2.9: overlapping part, considered to be relevant) occupies a large percentage of the period relative to the document D_f. In a similar way, the *overlapping* score [2.10] evaluates the ratio of the period relative to query Q occupied by the common period O. The higher this score, the more significant the corresponding document is. Finally, the *distance* score [2.10] evaluates the distance of the centroids of the period relative to query Q and that of the common period O. The closer these centroids are to each other, the more relevant the document fragment D_f is.

$$
\begin{aligned}
Precision(D_f, Q) &= \frac{O}{D_f} \\
Overlapping(D_f, Q) &= \frac{O}{Q} \\
Distance(D_f, Q) &= \frac{d}{D}
\end{aligned}
\qquad [2.10]
$$

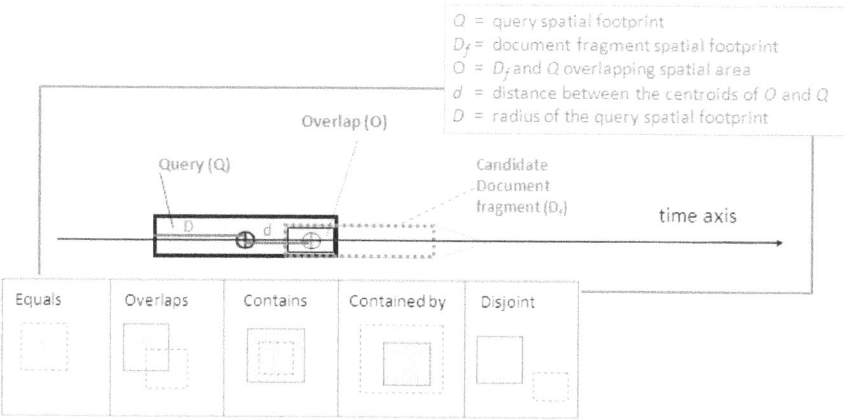

Figure 2.9. *Elements of measure of temporal similarity*

2.4.6. *Evaluation and Discussion*

We evaluate the effectiveness of the indexing performed by the PIV process flow on the one hand, and the quality of the IR supported by the PIV IRS on the other hand.

2.4.6.1. *Evaluation of the PIV indexing process flow*

The evaluation of the spatial indexing process flow [SAL 07c] has been performed on a sample composed of 10 books (travelogues), which have been scanned and gone through optical character recognition (OCR). The PIV process flow has detected 9,835 SFs. At the same time, in the manner of the CLEF[28] evaluation campaigns, a manual extraction and indexing of the SFs has been carried out in order to build a set of reference SFs.

To evaluate the effectiveness of the indexing, we have chosen to use measures of recall and precision, on the basis of a manual annotation of 10 pages for each of the 10 volumes. Evaluation reveals that the recall is equal to 0.49 and the precision is 0.73. The low rate of recall can be explained, after analysis of the non-recognized SFs, by the lack of geographic resources (for the validation and interpretation of these SFs), then, with a lesser importance, by the lack of spatial indicators in the glossaries, the orthographic variations and the problems in OCRization. The errors in precision are mainly due to homonyms or old names unavailable in our resources. These test results are detailed in [LES 07], which gives an analysis of the results of each step in the indexing process. The precision of the PIV SF extraction process flow is lower than that of the SPIRIT system presented in [CLO 05]: this system features a recall of 0.69 and a precision of 0.78, it indexes SFs of hotel, restaurant, street, postal code and commune type and uses comprehensive corresponding geographic resources for the validation phase.

The evaluation of the temporal indexing process flow [LEP 07] was carried out on a smaller sample size composed of text extracted from these same books. The PIV process flow detected 540 TFs. Evaluation reveals that the recall is equal to 0.91 and the precision is 0.97. These strong results need to be put into perspective: they can be explained by the fact that the implemented grammars have been defined from the preliminary study of a large part of the sample, which then served for the evaluation.

2.4.6.2. *Evaluation of the PIV IRS*

Experiments relative to the spatial and temporal IR supported by the PIV prototype are described in [SAL 07a], [SAL 07b] and [PAL 10a].

The spatial IR experiment focuses on a sample of texts extracted from these same books, composed of 1,019 paragraphs, which corresponds to 1,028 SFs (902 ASFs and 126 RSFs). The protocol contains 40 queries: 15 queries focus on ASFs (five with small spatial range such as "pass", five with intermediate range such as "commune" and five of large range such as "region"); 25 focus on RSFs (five for each type: adjacency, inclusion, orientation, distance and union). Three people have conducted a *pooling*-type assessment phase described in [PAL 10a]. The evaluation of these results of PIV spatial IR gives an MAP of 0.62 higher than that of the SPIRIT GIR system, which is equal to 0.40 [PUR 07]. Let us note that the SPIRIT system used 38 queries also referring to RSFs with relationships of adjacency,

28 Cross-Language Evaluation Forum (CLEF) – /http://www.clef-campaign.org/.

inclusion and orientation. On the other hand, the process of extraction and indexing of this system (see Table 2.1) only manages ASFs without taking into account the possible associated spatial relationships. The difference of IR precision between these two systems can be explained, in part, by this observation. It must also be noted that the document collection used for the evaluation of the PIV prototype is smaller in size and more localized (limited to the Pyrénées mountains) than that implemented for the SPIRIT system evaluation. This gives an advantage to the results of the PIV system in terms of precision. Moreover, let us note that during the evaluation of the SPIRIT system, Purves *et al.* [PUR 07] show the contribution of a spatial IR in complement to a thematic IR for answering these queries containing spatial and thematic criteria.

The experiment relative to the temporal IR, presented in [PAL 10a], focuses on the entirety of the 10 books, in other words 5,645 paragraphs (1,702 TFs: 1,561 ATFs and 141 RTFs). The protocol contains 35 queries: 15 queries focus on ATFs (five with small temporal range such as "day", five with intermediate range such as "month" and five with large range such as "year"); 20 focus on RTFs (five for each type: adjacency, inclusion, orientation and interval). Three people have conducted a *pooling*-type assessment phase described in [PAL 10a]. The evaluation of these results of PIV temporal IR gives a mean average precison (MAP) of 0.93. It was not possible to compare this to other systems such as Mondeca (see Table 2.1), which, to our knowledge, has not yet published test results of this type.

2.5. Summary

In this chapter, we have detailed our contributions concerning the spatial and temporal information extraction, interpretation, indexing and retrieval from texts. The main objective of these processes is the creation of indexes supporting spatial and temporal IR scenarios in textual document corpora. The PIV prototype implements these propositions.

2.5.1. *Contributions*

The PIV prototype supports four main modules dedicated, respectively, to the extraction, interpretation, indexing and retrieval of spatial and temporal information. The first two modules propose linguistic processes for the recognition, interpretation and symbolic annotation of information. Two core models, respectively, spatial and temporal, support this recognition, interpretation and representation, of toponymic and calendar references but also of spatial and temporal relationships. The two other modules are dedicated to the spatial and temporal information indexing and retrieval. Thus, in this IR context, the symbolic annotations are transformed into numeric representations organized into indexes essential to the spatial and temporal IR models [GAI 08, LEP 07]. The PIV prototype shows the feasibility and usefulness of such specialized IRSs.

The ad hoc spatial and temporal models that we propose are close to the SpatialML and TimeML models. The transition of one representation to another is made possible by model transformation, without the possibility, however, of associating numeric representations with the RSFs and RTFs in SpatialML and TimeML. We have made the choice of working with ad hoc models since, on the one hand, we conduct a simplified annotation in the framework of our process flows and, on the other, we do not have the need to exchange representations. The spatial and temporal extraction, interpretation and indexing processes, as well as the spatial and temporal IR models, have been implemented in an empirical way on textual samples extracted from our corpus. These indexing process flows and the corresponding IRSs have been tested. Like the work described in [PUR 07], our experiments [SAL 07a, LEP 07] show that the specialized PIV IRS gives better results than a classic thematic IRS for an IR containing only spatial or temporal criteria. In [SAL 07a], we also show that a "rough" coupling of spatial and thematic IRSs gives better results than a unique thematic IRS for an IR containing spatial and thematic criteria.

This work enabled us to develop different prototype versions each corresponding to contributions to the projects described in Chapter 1:

– Web services of spatial and temporal processing of textual documents: for the annotation of archival documents in the GEOTOPIA (see Figures 2.10 and 2.11) project and for the annotation of hiking reports in the GEOCIME project;

– process flows aiming at the extraction of geographic terms candidate for the enrichment of ontologies in the GEONTO project;

– specialized spatial and temporal IRSs dedicated to information indexing and retrieval in travelogues for the PIV project and in hiking itineraries for the GEOCIME project.

Let us also mention two other prototypes. On the one hand, a first extension of PIV is dedicated to spatial indexing by patterns [LES 07]. The corresponding prototype implements a method of indexing by classification guided by the evaluation of spatial characteristics specific to the SFs of indexed textual units. These characteristics allow the grouping of SFs and the association of *itinerary*, *local description*, *point of view*, and *comparison of places* patterns with documentary units with as diverse ranges as the paragraph, the section, the chapter, etc. The IR scenarios can now be different and the retrieved results no longer contain only paragraphs but also document fragments of different structure and size. On the other hand, the second prototype aims at the automatic interpretation of itineraries [LOU 08a]. This prototype, called PIIR, takes as the input a rough text related to a travel account and gives an interpretation in XML format of the steps corresponding to the itinerary described in the account as the output.

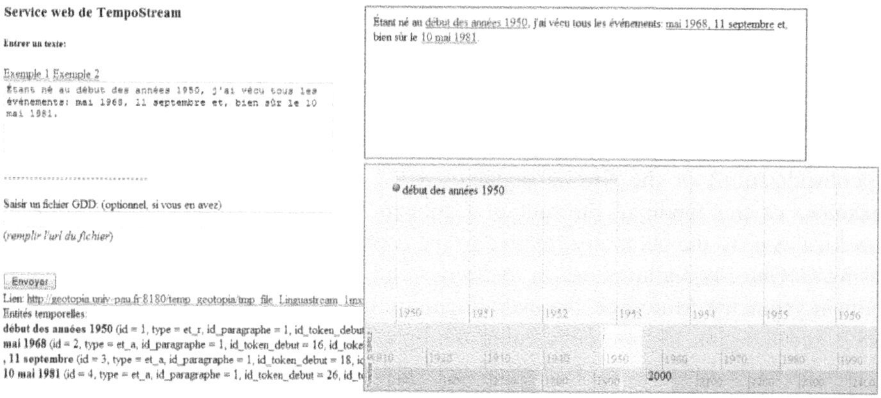

Figure 2.10. *Illustration of the temporal information annotation service intended for French texts – "Born in the early 1950s, I lived all well-known events: May 1968, September 11 and, of course, May 10, 1981"*

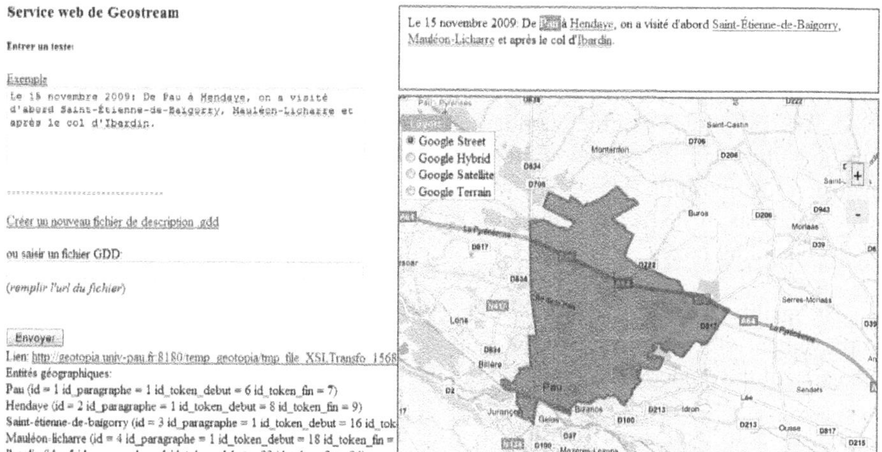

Figure 2.11. *Illustration of the spatial information annotation service intended for French texts – "November 15, 2009: from Pau to Hendaye, we first visited Saint-Etienne-de-Baigorry, Mauléon-Licharre and then the pass of Ibardin"*

2.5.2. *Perspectives*

The successive versions of PIV prototypes integrate notable evolutions of the process flow shown in Figure 2.6. We mention two evolutions that are of particular interest.

First, the integration of the geographic ontology GéOnto (see project GEONTO, Chapter 1) in the *semantic processing* stage has allowed an improvement of the quality of the classification of SFs and, consequently, of their retrieval in the geographic resources followed by the calculation of the corresponding numeric representations, in the next stage. The improvement of the interpretation quality (geolocalization) of the SFs is equal to 85%. This is easily explained with the example of the toponym "torrent of Cauterets", which, in the standard flow, is geolocalized by the geometry relative to the *commune* of Cauterets and, in the flow using GéOnto, is classified as an *hydronym* and geolocalized by the corresponding stream geometry. Moreover, the overall gain in time of the whole process is equal to 28% and the gain in time concerning the operation of geolocalization of an SF is equal to 84%. Indeed, the typing of the entities allows us to directly query the corresponding resource/table in the IGN gazetteer. This has a double effect: respecting the meaning attached to the corresponding toponym and increasing the parsing speed of the resource.

Another consequent evolution concerns the utilization of geometric primitives in order to refine the RSF representations initially approximated by the bounding boxes. We localize it in the continuity of the propositions made by Malandain [MAL 00] and by Fu *et al.* [FU 05] for the interpretation of information units extracted from textual or iconographic documents. Thus, our work described in [LES 07] and [SAL 08] proposes specific configurable functions taking advantage of GIS functionalities and fuzzy notions for new spatial representations of adjacency and inclusion relationships. These new algorithms improve the precision of the numeric representations of RSFs and reduce the problem of noise due to the approximations by the bounding boxes.

The quality of the spatial PIV IRS depends largely on these pieces of work on the improvement of ES typing with the help of ontological resources as well as on the precision of the numeric representation of the RSFs.

Let us note that, as in the spatial case, the creation followed by the usage of praxonymic temporal resources (historical events, diseases and cultural events) relative to the 18th and 19th Centuries and to the Pyrénées mountains could allow the improvement of the temporal PIV IRS.

All these orientations are at the core of the concerns and work of the T2I team, of the laboratory of LIUPPA, concerning spatial and temporal informations in textual documents. The propositions targeting the creation of a multicriteria IR system combining the spatial, temporal and thematic dimensions are based on this first foundation of results.

As we have just emphasized, the approaches presented in this section are better suited than those supported by classic IRSs for the retrieval of documents based on spatial or temporal criteria. However, for the retrieval of documents based on spatial, temporal and thematic criteria at the same time with a strong expressiveness such as “Pyrénées mountains but not those of Gavarnie, in the 19th Century, if possible, unrelated to ascents”, we would need to study different processes of IRS combination and their possible biases.

Thus, beyond these different versions of the PIV IRS, dedicated to the spatial and temporal information contained within textual documents, we will, in Chapter 3, design and implement a multicriteria IRS combining a spatial IRS, a temporal IRS and a thematic IRS in order to meet needs where the search criteria focus on these dimensions with an expressiveness that will allow us to associate constraints with each search criterion.

3

Multicriteria Information Retrieval in Textual Corpora

3.1. Introduction

Although regular search engines already retrieve good results for keyword-based searches, Kanhabua and Nørvåg [KAN 08] observed that, on specific corpora, the precision of geographic retrieval is weak. Indeed, users spend a lot of time exploring the retrieved documents in order to keep only the documents which satisfy their needs. For example, within a larger query, the temporal expression "the 1810s" submitted to a classic search engine leads to the retrieval of documents containing "1810" and not "1811", "1812", etc. Similarly, the spatial expression "around Anglet" submitted to a classic search engine leads to the retrieval of results containing "Anglet" and not "the gulf of Chiberta", "the beach of Cavaliers", "Bayonne", "Biarritz", etc.

A means of enhancing the efficiency of search engines is then to take into account not only the thematic aspects, but also the spatial and temporal aspects. In Chapter 2, we dealt with spatial and temporal dimensions as privileged entry points to the texts. The PIV project we presented is mainly based on the development of process flows dedicated to the indexing and retrieval of spatial and temporal information. A prototype, corresponding to each process flow, extracts and indexes the information from textual documents and proposes a search engine which, based on spatial [GAI 08] or temporal [LEP 07] criteria, returns paragraphs of documents.

Our objective now consists of proposing a better approach combining standard IR services with specific ones dedicated to the spatial IR and temporal IR. We propose a unified model representing each of the three dimensions and aggregating the results from different specialized IRSs which we implement in a new version of the PIV prototype. To validate our multi-dimensional approach, we also propose an evaluation framework for geographic IRSs (GIRSs). The proposed evaluation framework is destined to evaluate spatial, temporal and thematic IRSs or any combination of such IRSs. It provides a test collection in French language adapted to the GIR and, in the medium term, could lead to the proposition of a specific task in the framework of an evaluation campaign such as GEOCLEF [GEY 05].

The contributions described in this chapter are the result of study conducted on the framework of the T2I team of the laboratory of LIUPPA and, in particular, of the CIDRI, PIV^2 and PIV^3 projects. These projects are detailed in [PAL 10d], [PAL 11], [PAL 12a] and [SAL 11] and in the thesis of Palacio [PAL 10a]. They are also the result of collaborations with Guillaume Cabanac and Gilles Hubert, researchers at the IRIT involved in IR activities: together, we have, in particular, set up a GIRS evaluation framework which we have applied to the multicriteria GIR prototype PIV^3 [PAL 10c, CAB 11].

This chapter is organized into six sections. Following this introduction, section 3.2 recalls the research questions raised in Chapter 1, which will be studied here. Section 3.3 presents the propositions found in published literature dealing with the issues of criterion standardization, on the one hand, and criterion combination, on the other. Section 3.4 details the multicriteria GIRS PIV^3 that integrates our propositions for indexing by tiling and the multicriteria IR. Section 3.5 describes the GIRS evaluation framework we have designed as well as the results of the evaluation of the GIRS PIV^3. The summary of our contributions and perspectives are presented in section 3.6.

3.2. Review of challenges, hypotheses and research objectives

Few research studies are interested in the combination of classic IR services with specific services dedicated to spatial IR and temporal IR. We can mention Alonso *et al.* [ALO 11] who propose the use of the temporal information annotated in texts in order to constrain a search engine: a query is now composed of a *textual* part completed with a *temporal* part. Alonso *et al.* illustrate this proposal with the example of the query "world war" temporally constrained by the criterion "1939–1945". The search thus targets the documents dealing with the Second World War. Alonso *et al.* consider that a query combining *textual* and *temporal* search criteria has to be processed in the following way: the higher the level of satisfaction of these criteria, the higher the corresponding score of the document. They, however, raise three questions:

– How should a combined score coming from *textual* and *temporal* search criteria be calculated reasonably (side effects are probable)?

– Does a document in which the *textual* and *temporal* elements of correspondence are distant have to be penalized?

– How should a document, which entirely satisfies *textual* criteria but little or no *temporal* criteria, be weighted and vice versa?

We place our issues (see Chapter 1) in this same framework for reflection. Our objective is to combine a classic IRS with spatial and temporal IRSs. The heterogeneity of the corresponding models of information representation and retrieval does not allow us to directly consider the combination of such systems. For example, the aggregation of results coming from different models of IR raises problems. Thus, the first question dealt with in this chapter is: *What core models of information representation and retrieval should be proposed in order to prepare for the combination?* which proposes a preparatory phase in advance of the result combination phase itself. Consequently, a phase of model standardization will be the basis of the controlled combination approaches.

Finally, the combination has to be based on aggregation operators adapted to the context. Let us not forget the need for expressiveness mentioned earlier. This need is partially covered by the operators proper to each dimension. It nevertheless has to be completed by a finer formulation of each criterion: is it mandatory, is it associated with a degree of preference, is it a rejection criterion? The question: *What advanced aggregation operators should be proposed and implemented for a multicriteria IR combining several IRSs?* will thus be dealt with in a second stage.

The retained working hypotheses are the following:

– The standardization of data representation is adapted to spatial and temporal information.

– The standardization associated with a well-tried IR model does not involve a loss of precision with respect to the initially dedicated IRSs.

– A "classic" coupling (such as an arithmetic mean, for example) of spatial, temporal and thematic IRSs gives better results than a thematic IRS alone or two-by-two pairings of spatial, temporal and thematic IRSs.

– An "advanced" coupling (offering more power of expression to the user) of spatial, temporal and thematic IRSs gives better results than a "classic" coupling.

The research objectives corresponding to the announced issues aim mainly at the design and implementation of:

– a model of standardization of indexed data representation;

– a model of multicriteria IR.

These hypotheses are some of the avenues of study that we have followed in order to implement a GIR fully exploiting the spatial, temporal and thematic dimensions.

3.3. Standardization and combination of criteria: literature review

The combination of criteria is an issue that mainly appears in the fields of IR and decision support systems (DSSs). To avoid a number of possible biases during the combination phase, a preliminary phase of criteria standardization is generally proposed in the approaches relative to each of these fields. Indeed, the heterogeneity of the models of data representation and criterion evaluation involved in the combination has to be taken into consideration.

3.3.1. *Criterion standardization*

We are interested in the approaches to standardization in the fields of IR, multimedia IR, DSSs and GIR.

3.3.1.1. *Normalization and IR*

Normalization consists of bounding the relevance score of a document calculated with respect to the given query. In IR, these scores are mostly bounded between 0 and 1 [KOW 97]. If a query is multicriteria, each of the criteria has to be bounded in order to be combined later with the others. In a keyword search, a user can specify several criteria, in other words several possibly constrained words. The IRS will then perform a search for each criterion and then build a result list composed of ranked documents. A normalized score (average of the scores, for example) is associated with each document. In this particular case, given that the data representation model is unique and the same IRS is used to process each of the criteria, there is no problem of compatibility between the obtained results.

3.3.1.2. *Generalization and multimedia IR*

Generalization corresponds here to an approach of simplification aiming at the reduction of a very large quantity of details. Several approaches allow us to achieve this simplification. The first is identification: based on multimedia data, the concepts [WOL 92] (a concept corresponds to a meaningful piece of information, a category such as car, person and house) present in the document have to be determined. The second is regrouping/focusing: based on multimedia data, the information detected in the document has to be regrouped in order to be associated with more general representation frameworks. For example, the so-called administrative hierarchy of places (communes, counties, regions, etc.) can be used to simplify the representation of spatial information [LIE 07, MAR 08b].

The identification approach to generalization is used to determine the concepts contained in a document. It is necessary beforehand to define the list of suiting concepts and characterize them. It is possible to apply the generalization approach to different images (25 images per second) and to the audio track of a video in order to detect the concepts on each media. Ayache *et al.* [AYA 06] do not use the same base of concepts for the two document types: they propose 15 concepts for images and 100 or so for audio documents.

The regrouping approach to generalization consists of gathering several pieces of information of the same kind. This regrouping does not necessarily lead to meaningful information as with the identification approach, which does not allow us to build reference bases such as concepts. For example, the truncation in IR represents terms with truncated terms (thus, "advantages" and "advantaging" can be regrouped into "advantag"). Pham *et al.* [PHA 07] propose the application of this approach to images by splitting the image with a regular grid to obtain a set of cells (called *patches*). An image can thus be considered a bag of *patches* (just as a text is considered a bag of terms). Pham *et al.* call these *patches visterms* (visual terms) by analogy with the terms contained in texts (which they call *texterms*). When splitting an image in *visterms* using a regular grid, the same *visterm* can appear several times; thus, its raw frequency determines its weight just like for the terms. Pham *et al.* also propose a segmentation of the images in regions, which consists of performing a splitting guided by the content (colors, for example). Nevertheless, this approach does not allow the semantic identification of the constructed regions.

3.3.1.3. *Standardization and decision support systems*

Mainly used in multicriteria decision support, standardization consists, just like normalization, of bringing the score of the different criteria along the same interval of results ([0;1], in general). Nevertheless, the main difference with normalization is that standardization is applied to non-numeric evaluations [MAL 03, LAB 03]. For example, each criterion is evaluated following three categories (strong interest, average interest and weak interest) in [MAL 03] and following six categories (very weak, weak, average, strong, very strong and exceptional) in [LAB 03]. Moreover, standardization demands that, for each criterion, the categories be compared in pairs. In [MAL 03], each pair is labeled from 1 to 9 by the user, which indicates whether the two categories are of equal importance (1) or one of them is a lot more important than the other (9). Once this pair-comparison matrix has been established, the system generates the corresponding standardized numeric values [SAA 80, MAL 03]. An example of a standardization process with an added comparison phase of category pairs is illustrated in [PAL 10a]. It is a study in the automotive domain. The categories corresponding to the color criteria are "ideal" for blue and red, "acceptable" for green and orange and "borderline" for white and black. The comparison phase of these category pairs determines the values of 1 for "ideal", 0.75 for "acceptable" and 0.3 for "borderline": the green and orange cars will thus have a weight of 0.75 for the color criteria.

DSSs have to conform to the preferences of the user. A preliminary system adjustment phase is therefore necessary. This setting phase is dependent on the user: for each criterion, the user is encouraged to classify the possible results in categories/levels of interest and also to define the difference felt between each pair of categories.

3.3.1.4. *Spatial grouping and GIR*

Lieberman *et al.* [LIE 07] and Martins *et al.* [MAR 08b] tackle the problem of combining the place references given in a text in order to find the encompassing geographic scope that the document discusses as a whole. They propose the standardization of spatial information by mainly searching for the "focus" of the document, in other words the main area which it relates to. The advantage of this is the simplification of the spatial search, given that a single piece of spatial information is associated with each document. The administrative divisions being hierarchical (the world is divided into continents, followed by countries, regions, cities, etc.) and meaningful, Lieberman *et al.* [LIE 07] and Martins *et al.* [MAR 08b] implement this type of standardization. It is thus possible to regroup several cities within the same region, for example.

By applying a similar approach to the temporal and thematic dimensions, in other words by obtaining the spatial, temporal and thematic scopes of the document, the combinations would apply in a logical way. Indeed, each criterion subject to a generalization-type standardization, the criteria would then be consistent. Nevertheless, as Lieberman *et al.* [LIE 07] explain, unless the inaccuracy of the results is very strongly increased, it is very difficult to obtain a single focus if the document contains very distant spatial information. Thus, for example, for a document evoking the city of Biarritz (in Southwestern France) and the city of Lille (in the North of France), the process of generalization will propose France. This form of grouping can, therefore, lead to an excessive generalization [LIE 07].

3.3.2. *Combination of criteria*

We have studied the approaches dedicated to the standardization of criteria in fields such as IR, multimedia IR, DSSs and GIR. We will now show interest in the approaches to the combination of criteria in these same fields. Nevertheless, it has to be specified that, depending on the field, the term used is not *combination* because even if the aim is to unite several criteria, the context, the number of criteria and the methods can be very different. It is called *criteria fusion* for multimedia IR and *criteria aggregation* for DSSs. *Criteria fusion*, on the one hand, allows us to gather sets of information extracted from different documents (e.g. images and texts) or from different parts of the same document (e.g. a video can be broken down into a set of images and a soundtrack). *Criteria aggregation*, on the other hand, is used to help a user make a decision corresponding to the best compromise (all the criteria can rarely be satisfied). Concerning the GIR, the modes of combination are generally of *filtering* type: they consist of successively querying the IRSs dedicated to each of the dimensions followed by performing the intersection of the results. Finally, the multicriteria IR integrates the processes of *aggregation* which, depending on the case, allow a criterion to compensate for one or more other non-represented criteria.

3.3.2.1. *Fusion and multimedia IR*

Generalization, as explained in the previous section, enables *rough* information extracted from a document into concepts [WOL 92]. An example of fusion, applied to a video, is illustrated in [PAL 10a]. In a preliminary generalization step, the images (colors, textures, forms, etc.) and the audio track (sounds, speech, etc.) are analyzed and higher level concepts are extracted (e.g. "car" and "sport"). Fusion then consists of gathering information of the same nature. For a video, the concepts extracted from the images and from the audio track are fused [AYA 06]. For an image on a Web page, for example, the operation aims to fuse information extracted from the image and from the text it is associated with [PHA 07]. It has to be noted that this fusion is done in an equitable way: when an image and text are fused, none of them benefits. Indeed, in the different studies mentioned, none weigh the different types of data used. This fusion approach, therefore, allows us to generate new indexes corresponding to a higher level of abstraction qualified as semantic or conceptual.

3.3.2.2. *Aggregation and decision support systems*

DSSs have been used for approximately 40 years to facilitate decision-making [POW 97]. *Reality* is multi-dimensional [ZEL 82]; consequently, the criteria (also called constraints) to take into account for decision-making are numerous [BOU 93]. The objective of a multicriteria decision support is to propose to its user scenarios best satisfying these previously defined criteria in the search of a solution. It has to be noted that the expressed criteria cannot all necessarily be respected in the proposed scenarios. Thus, criterion aggregation is used in order to find the best compromises. These approaches are based on a strong interaction with the user; he or she has to determine the importance of each criterion by assigning a weight to it and modulate the desired compensation between criteria.

More precisely, every multicriteria decision problem can be formulated by the model "A, A/F, E" [MAR 99]. A is the set of alternatives, in other words the possible choices. A/F is the set of attributes or features (*Attribute/Feature*), in other words the constraints expressed by the user. E is the set of the evaluations of the alternatives' performance according to each criterion; the extent to which the alternative fits the criteria has to be measured. Let us take the example of the selection of a plot of land for the construction of a new housing project. The set of alternatives (A) corresponds to the different plots of land offered for sale: $plot_1, \ldots, plot_n$. The set of features (A/F) determines the quality of such a plot in a property context: 1–low noise pollution, 2–unfloodable zone, 3–affordable prices, 4–immediate availability, etc. The evaluation of the performance of each of the available plots with respect to these criteria generally results in the attribution of scores.

Different approaches to criterion aggregation are presented and illustrated in [PAL 10a]. The evaluation of the criteria can be based on a type of Boolean calculation or proportional quantitative calculation. Such approaches, for which all the criteria are of equal importance, allow us an alternative less satisfying criteria to be ranked equally well or better because of a higher score of the satisfied criteria [YAG 88, MAL 03]. Consequently, preferences can be associated with the criteria. These allow us to promote certain criteria and influence the mode of compensation [MAL 03, BÜY 10]. Let us note that the obligation to satisfy a given criterion is not taken into account.

In parallel, there are numerous studies which mainly look to attenuate the compensation created by the definition of preferences and the quantification of criteria-alternative couples. The ordered weighted averaging (OWA) method [YAG 88] allows us to specify the level of compensation between criteria. For example, the OWA "*orand*" operator proposes sorting the criteria by order of decreasing scores and defining the weight for each rank instead of each criterion, the total weight not exceeding one (an example of OWA is described in [PAL 10a]). This method enables us, by weighting, to parametrize an approach by total compensation, medium compensation or without compensation. Pereira *et al.* [DA 09, DA 12] propose a method allowing the use of four criteria to evaluate the relevance of a document: *aboutness*, *coverage*, *appropriateness* and *reliability*. Besides associating a weight with each of these criteria, they also associate an order of priority with them. For example, when the aboutness is the criterion of priority number 1 and it is not satisfied, the other criteria are not examined. In a more general way, whatever the criteria, the order in which they are expressed corresponds to the level of importance the user wishes to assign to them; Pereira *et al.* [DA 09, DA 12] emphasize the particularly intuitive characteristics of this approach.

All these approaches consider that the different criteria are independent. The weighted arithmetic average proposed in most of these studies does not allow us to take into account the possible relations existing between criteria. Several published studies propose the use of methods such as the Choquet integral [BÜY 10, LAB 03, HAM 09]. For example, Büyüközkan and Ruan [BÜY 10] explain that, in the framework of risk evaluation in software development, some risks (criteria) are linked. Instead of associating a weight with a criterion, the Choquet integral allows us to assign weights to pairs of criteria. As explained by Labreuche and Grabisch [LAB 03], it is necessary to identify such links and measure them, which is generally difficult for a non-specialist.

3.3.2.3. *Filtering and GIR*

In Chapter 2, we have defined geographic information composed of three dimensions: spatial, temporal and thematic. We have presented various existing GIRSs and noted that many systems were only interested in two of these dimensions (generally spatial and thematic). Palacio [PAL 10a] proposes regrouping the approaches to result combination implemented by these GIRSs in three categories: filtering (sequential or parallel), linear combination and projection. Let us note that the filtering approaches are the most widely employed. A presentation of these categories, illustrated with examples, is proposed in [PAL 10a].

The approach corresponding to sequential filtering consists of performing a first retrieval based on criteria of a given dimension: the result is a set of ranked relevant documents. Then, a new retrieval, based on criteria relative to another dimension, is carried out on this set of results. Finally, a last retrieval concerns the third geographic dimension. The first two retrievals allow us to reduce the set of resulting documents (filtering approach), and the last retrieval allows us to obtain a set of ranked results with respect to the criteria relative to the last processed dimension. Thus, sequential filtering allows us not to combine the three dimensions directly. The STEWARD [LIE 07] and SINAI [GAR 09] systems implement such an approach for a thematic and spatial filtering. Lieberman *et al.* [LIE 07] argue in favor of a contextual configuration of the order of use of dimensions in the process of sequential filtering.

The approach corresponding to parallel filtering consists of querying the document repository with respect to each of the dimensions of the query separately, then performing the intersection of the results. Thus, the query is subjected to three processes proper to each geographic dimension. For every dimension, a set of results is retrieved. Only the documents present in the three sets are kept. No ranking is performed. However, in the set of results obtained, some documents are evidently more relevant than others and should appear on top of the list. This approach does not favor any dimension as does sequential filtering, for example. The SPIRIT [VAI 05], GRID [VAL 06], GEOTRACKER [CHE 07], WatWasWaar [LIB 08], CITER [PFO 09] and Document Trajectory [STR 10] systems implement parallel filtering.

Other approaches to combination are applied to GIR. Linear combination allows us to combine scores from several IR systems [VOG 99], the objective being to obtain a unique score. For example, the arithmetic average corresponds to a linear combination associating an equal weight with each dimension: GEOSEM [BIL 07] calculates the average of the scores of the spatial, temporal and thematic IRSs. The GEOSEM [BIL 03], DIGMAP [MAR 07], PIV [GAI 08], GEOOREKA [BUS 09b] and Local Search [BRI 10] systems implement linear combination.

A final approach to combination called "scattered ranking" [VAN 05] targets the reorganization of the retrieved documents in order to increase their diversity. The aim is to relegate a part of the documents with similar scores further in the ranking in order to increase the variety of the results or to regroup them as proposed by Google. The approach consists of spreading the documents (results) on a frame of n dimensions, those closest to the origin of the frame are the most relevant. Then, the points of the frame (results) are compared in pairs, and if a result is too close to another it will be projected further away in the classification (frame). Kreveld *et al.* [VAN 05] pick up this approach for the spatial and thematic dimensions, whereas Purves *et al.* [PUR 07] test it on the SPIRIT system. This approach does not allow us to modulate the combination (by favoring a criterion, for example), given that ranking of results is recalculated dynamically.

3.3.2.4. *Aggregation and multicriteria IR*

In IR, Kelly and Fu [KEL 07] have shown a strong relationship between query extension and performance. Similarly, in multicriteria IR, Croft and Harabagiu [CRO 00] have shown the interest of approaches combining different strategies of information representation and retrieval in textual content as techniques improving the effectiveness of IR. They emphasize three categories of approach: the combination of different representations of the corpus before the IR algorithms, the combination of different IR algorithms and the combination of results from different IR algorithms. The aggregation models based on the relevancy scores proposed by Fox and Shaw [FOX 93], as well as by Miriam Fernandez *et al.* [FER 06], are found in the third category. They propose a normalization of the scores before implementing the aggregation phase. Farah and Vanderpooten [FAR 08] define result aggregation as a process of ranking documents combining the scores ("retrieval status values" or RSV) obtained for each search criterion. Figure 3.1 illustrates the principles of multicriteria IR. A multicriteria query translates the user need. The search engine analyzes the corpus in such a way as to find correspondences between the contents of the documents and the search criteria. A list of results L_j, containing documents and their relevance scores, is created for each criterion $Criterion_j$. Then, the search engine aggregates these lists according to an aggregation function f and produces the final result list L which is presented to the user.

Farah and Vanderpooten [FAR 08] define three classes of aggregation functions applying the principles illustrated in Figure 3.1:

– *Totally compensatory aggregation:* This consists of computing a score from a set of scores rsv_{ji} assigned to a document d_i for each criterion of the query, by applying a function f as the weighted sum, for example. A weak score for one criterion can perhaps be balanced by one or more higher scores obtained for other criteria. So this kind of approach does not handle any mandatory criterion whose satisfaction would be required for any result. Prominent compensatory aggregation functions are Borda Count [DEB 81], CombMAX, CombMNZ and CombSUM [FOX 93].

– *Non-compensatory aggregation:* In this case, aggregation is mainly based on the value of one criterion, such us the weakest score rsv_{ji} assigned to the document d_i. The remaining criteria are only used to discriminate between documents of similar scores. The main disadvantage of this class of aggregation functions is that a large part of the criteria (and the corresponding scores) are ignored or play a minor role. CombMIN [FOX 93] and Prioritized And [DA 09, DA 12] are non-compensatory aggregation functions.

– *Partially compensatory aggregation:* Farah and Vanderpooten [FAR 08] propose an approach to result combination that takes into account the underlying vagueness of the query formulation process. Similarly, Pereira *et al.* [DA 09, DA 12] propose the "Prioritized Scoring Model" (PSM) aggregation function that assigns the highest weight to the first criterion and decreasing weights to the following criteria.

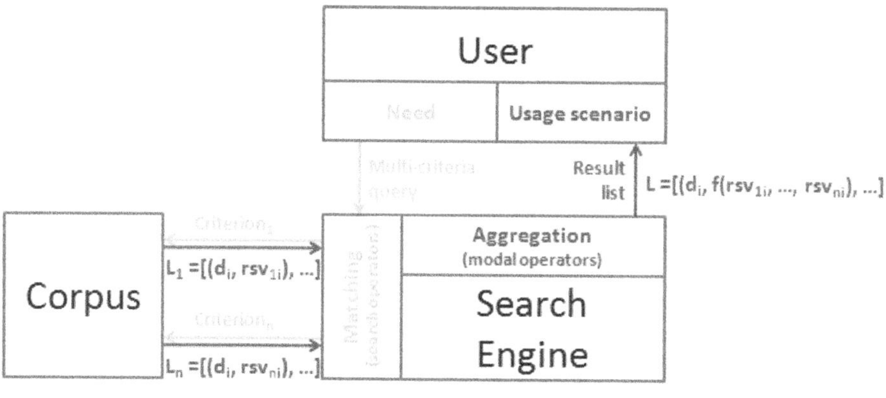

Figure 3.1. *The principles of multicriteria IR [PAL 12a]*

Consequently, we rank GIRSs based on approaches of linear combination in the category of totally compensatory aggregation functions. However, let us note that the results proposed by these systems could be biased: indeed, they aggregate results from dedicated IRSs characterized by specific models of data representation and relevance score calculation.

Similarly, we associate the GIRSs based on filtering approaches with the category of non-compensatory aggregation functions. These systems process sequentially, or in parallel, each dimension and calculate the intersection of the corresponding result sets. The final result is unranked or ranked following only one dimension.

To our knowledge, no GIRS implements partially balanced aggregation.

3.3.3. *Summary and positioning of a partially compensatory GIR*

Section 3.3 presents different studies concerning the standardization and combination of criteria in different fields, such as IR, multimedia IR, GIR and decision support. It has allowed us to grasp the elements in published literature dealing with our issue, which is the proposal of models and methods for the generalization of indexed data representation, on the one hand, and multicriteria IR, on the other.

From this study, it follows that the IR systems targeting several dimensions, such as GIRSs, usually federate mono-dimensional IRSs, each supporting their own indexes and matching operators dedicated to their field. GIRSs have, in general, done without the preliminary index standardization step, by implementing relatively simple combination approaches such as filtering of linear combination. Thus, we retain the approach proposed by Pham *et al.* [PHA 07] in multimedia IR, dealing with images in a way similar to texts, in other words using visual *lemmas* (*visterms*) and calculating weights relative to their raw frequency. This approach leads to homogeneous forms of representing different types of information. We propose representing spatial and temporal information according to an approach of standardization by spatial and temporal tiles followed by testing our first two working hypotheses:

– the generalization of data representations is adapted to spatial and temporal information;

– the generalization associated with a well-tried IR model does not lead to a loss of quality with respect to the initial dedicated IRSs.

We thus propose measuring the eventual loss of precision induced by generalization by comparing the ad hoc IR models, implemented in different versions of the PIV prototype, with the vectorial IR model applied to the new generalized indexes. We choose the vectorial model [SAL 71, SAL 75], well-tried and giving good results [BAE 99] and, although dedicated to terms, we apply it to spatial and temporal tiles. We study the interest of integrating the frequencies of tiles into the calculation of relevance scores by the vectorial model. Moreover, we observe the impact of the different range choices for spatial and calendar tiles on the precision of vectorial IR.

From the previous study, it can also be seen that the approaches of criteria combination in GIR use dedicated IRSs for each dimension, on the one hand, and do not allow any configuration of the combination on the other. We retain the approaches of multicriteria decision support whose flexibility allows us to specify, for each criterion, the level of compensation desired by the user [MAL 03, BÜY 10]. Similarly, the approaches of partially compensatory aggregation, proposed in multicriteria IR, authorize the weighting of each criterion according to the desired level of importance [FAR 08, DA 09, DA 12]. We propose adapting these approaches to design a partially compensatory GIR model which we have named CMRP ("criteria, matching operators, requirements, preferences" quadruplet for query aggregation): each search criterion (1) corresponds to one of the three spatial, temporal or thematic dimensions; (2) names a matching operator compatible with this dimension and, finally, (3) refers to a level of requirement or preference that will denote the importance of the criterion for the search. We study the interest of several types of IRS coupling, including the new CMRP approach that takes on an increased power of expression. Thus, we test our two other working hypotheses:

– A "classic" coupling (an arithmetic mean, for example) of spatial, temporal and thematic IRSs gives better results than a single thematic IRS or a two-by-two pairing of spatial, temporal and thematic IRSs.

– An "advanced" coupling (offering greater power of expression to the user) of spatial, temporal and thematic IRSs gives better results than a "classic" coupling.

We propose a new framework of evaluating GIRSs and measure the precision of a new PIV prototype implementing respectively *classic* and *advanced* (CMRP) couplings of spatial, temporal and thematic IRSs.

We present a standardization model of geographic information in spatial, temporal and thematic tiles in section 3.4.1. This involves the creation of indexes of a higher level of abstraction. We will then describe, in section 3.4.2, how the vectorial model is applied to such indexes and, finally, we propose a new aggregation model (CMRP) of the results of spatial, temporal and thematic IRs in section 3.4.3.

3.4. Proposition for indexing by tiling and multicriteria IR in textual corpora

We have chosen to deal with each dimension of the geographic information in a specific way. Before any combination, we have shown interest in standardizing the representations as well as the process flows relative to these data (see section 3.3). Our objective is therefore to create a generic approach to standardization that can be applied to each of the geographic dimensions.

3.4.1. *Standardization by tiling*

We complete the spatial and temporal indexing flow described in Figure 2.6 in section 2.4. A fourth stage, dedicated to the generalization of unitary indexes, is thus added to the process flow (Figure 3.2).

The fourth stage, thus, applies to unitary indexes in order to create a second level of index based on tiles. For the spatial dimension, for example, generalization corresponds to a regular or administrative (city, county, etc.) splitting of the area covered by the document repository followed by a projection of the spatial features (SFs) of the unitary index in this grid. Following the same principle, temporal standardization corresponds, for example, to a calendar grid on which the temporal features (TFs) of the unitary index are represented.

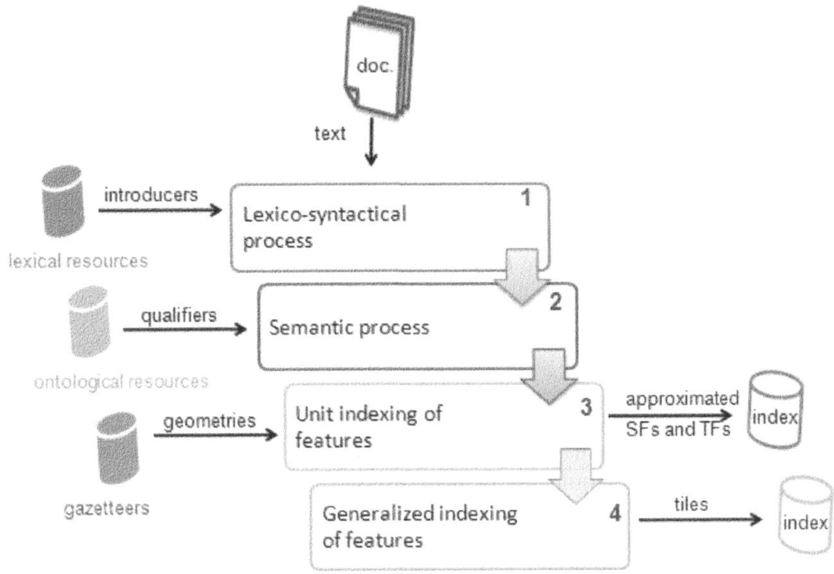

Figure 3.2. *A fourth stage in the PIV indexing flow*

This additional indexing stage presents a twofold interest. It allows us to bring the various spatial and temporal representations to a homogeneous representation supported by uniform spatial and temporal tile-based segmentations. Moreover, it allows the implementation of state-of-the-art models for computing IR relevance scores based on such spatial and temporal tiles given their frequency in documents. Such a generalization (i.e. standardization) leading to gathering geographic features into tiles causes some loss of accuracy [CHR 90]. This is balanced, however, by the introduction of tile frequency computation, as well as the computation of indexes of different scales (i.e. ranges). Thus, for example, the city zoning index (respectively county index) will be used if the query targets a city (respectively a larger area).

3.4.1.1. *Standardization by tiling: example*

Let us take the example of a text containing the sentence "I passed near Bayonne in early January 2001". The PIV process flows will analyze this text and create a first level of indexes associating a geometry with the SF "near Bayonne" (see $SF_{\#5}$,

Figure 3.3(a)) and an interval of time with the TF "in early January 2001" (see $TF_{\#9}$, Figure 3.3(b)).

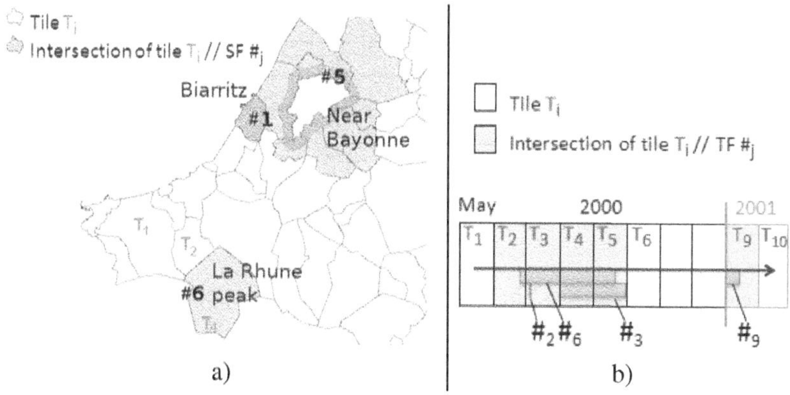

Figure 3.3. *Example of an application of tiling: a) spatial features on a communal administrative tiling and b) temporal features on a monthly calendar tiling – illustration taken and then adapted from [PAL 10d, PAL 11]*

Standardization by tiling will then be performed in two steps: the choice of the segmentation for each dimension and then the representation of the information of each first-level index in this tile-based grid. Thus, the spatial information needs a regular or administrative tiling (commune, county, etc.) of the zone covered by the document repository and a calculation of intersections of the SFs of the index and of this tiling. Similarly, temporal standardization needs a regular or calendar tiling (day, week, month, etc.) of the period covered by the document repository and a calculation of intersections of the TFs of the index and of this tiling.

Figure 3.3(a) illustrates the communal administrative tiling of the zone. The SF "near Bayonne" ($SF_{\#5}$) is in intersection with seven tiles (communes surrounding Bayonne). Similarly, Figure 3.3(b) illustrates the monthly calendar tiling of the period. The TF "in early January 2001" ($TF_{\#9}$) is included in the T_9 tile (month of January 2001).

The index excerpt presented in Table 3.1 illustrates the application of temporal tiling described in Figure 3.3(b): we can find the corresponding frequencies for the tiles $T_1, T_2, \ldots, T_n$. For example, $TF_{\#6}$ is in intersection with four tiles (T_2, T_3, T_4, T_5); the binary frequency relative to each of these tiles will be increased by 1. As for the tile T_3, it is in intersection with two TFs ($TF_{\#2}$ and $TF_{\#6}$); consequently, it will have a standard weight of 2. The proportional frequency takes into account the surface of coverage corresponding to the representation of a feature and to that of a tile. It also takes into account the scale of the feature with respect to that of the tile. For example, such a calculation will determine a proportional frequency of 1.08 for the application of $TF_{\#2}$ and $TF_{\#6}$ on the tile T_3.

id_t	List of id_{TF}	Cumulated binary frequency (TF//tile)	Cumulated proportional frequency (TF//tile)
T_1	[]	0	0
T_2	[$TF_{\#6}$]	1	0.21
T_3	[$TF_{\#2}$; $TF_{\#6}$]	2	1.08
T_4	[$TF_{\#3}$; $TF_{\#6}$]	2	1.98
...			

Table 3.1. *Example of a temporal tile-based index – taken and then adapted from [PAL 10d]*

These indexing strategies allow us to propose several indexes based on spatial and calendar tilings: spatial indexes whose segmentations are of district, city, township, county, region, country level and as well as calendar indexes whose segmentations are of day, week, month, season, year and century level. In a GIR context, these indexes then allow a search engine to browse the index most adapted to the spatial and calendar range of the query.

3.4.1.2. *Standardization by tiling: formalization of the approach*

This type of standardization consists of using a *tiling* to describe the spatial or temporal representations stored in the first-level indexes generated by the PIV process flows. This approach is a form of discretization, for example it associates a temporal object of the calendar domain with a temporal tile corresponding to a particular segmentation of the same space.

In a more formal way (equation [3.1]), a domain T included in the space $\mathbb{R}^n$ corresponds to the domain[1] O included in the space $\mathbb{R}^n$[2]. The domain O is composed of a set of objects $O_1, \ldots, O_p$ (the features of the first-level index), and the domain T is composed of the union of m subspaces (the tiles of the standardized index). For each subspace (T_i) of T in intersection with one or more objects (O_j) of O, we retain the number of intersections (N_{T_i}).

$$\begin{array}{l} O \subseteq \mathbb{R}^n \longrightarrow T \subseteq \mathbb{R}^n \\ O = \{O_1, O_2, O_3, \ldots, O_p\} \\ T = \bigcup_{i=1}^{m} T_i \\ N_{T_i} = Card(\{T_i \mid T_i \cap O_j \neq \varnothing, \forall j = 1, \ldots, p\}) \\ \text{with } Card(x) \text{ the cardinality of } x \end{array} \quad [3.1]$$

1 A domain is a finite or infinite set of values. It is represented by a list of elements or a necessary and sufficient condition of belonging – the domain of Booleans: {0,1}, the domain of the fingers of the hand: {thumb, index finger, middle finger, ring finger, pinky}, the calendar domain.

2 This superset refers to spaces of dimension 1, 2 or more.

This approach, described in [PAL 10a, PAL 11], therefore uses a specific existing index (first-level index) and generates a new standardized index (Figure 1.5, section 1.4.2, Chapter 1). Thus, to a set of one-dimensional representations, we apply a one-dimensional tiling and, to a set of two-dimensional representations, a two-dimensional tiling and so forth, until n dimensions. Let us note that we are here talking about spatial and temporal domains and the number of dimensions (1, 2 or more) necessary for the representation of information. For example (Figure 3.3), the specific temporal index references calendar representations viewable on the time line (one dimension). The corresponding standardized index contains tiles (months) materialized on the same time line. Similarly, the specific spatial index references geometric representations viewable on a map (two dimensions). The corresponding standardized index contains tiles (commune) materialized on the same map.

3.4.1.3. *Standardization by tiling: from the choice of tiling to the weighting of the tiles*

Indexing is composed of two steps: the establishment of the tiling and the weighting of the tiles. The first step thus consists of generating the different-range tilings. The second consists of assigning a weight to each tile according to its frequency in a document. We have identified two types of tilings: the "*regular*" tiling and the "*explicit*" tiling described and illustrated with examples in [PAL 09], [PAL 10d] and [PAL 10a].

– *Type of tiling: Regular* tiling is a tiling that consists of splitting the zone/period covered by the corpus into tiles of similar size. This approach is comparable to truncation. The size of the tiles and their borders are adjustable. This type of tiling is presented in [PHA 07] for *visterms*.

Explicit tiling consists of using an already defined and meaningful tiling. This approach is similar to lemmatization. This tiling is thus based on a splitting which we have qualified "*significant*" because it is built on human criteria (common sense).

Consequently, we associate *regular* tiling and *calendar* tiling with temporal information. *Calendar* tiling is an *explicit* tiling that consists of using the standard way of splitting: days, weeks, months, seasons, years, centuries, etc. This splitting allows the definition of several indexes corresponding to these different levels of calendar precision. Moreover, we associate *regular* tiling and *administrative* tiling with spatial information. *Administrative* tiling is an *explicit* tiling that consists of using the standard way of splitting: districts, cities, townships, counties, regions, countries, etc. This splitting allows the definition of several indexes corresponding to these different levels of spatial precision.

– *Weighting of tiles:* After having chosen the type of tiling, we can weigh the tiles by using the approaches based on the raw frequencies. To calculate the frequency of a tile, we propose two discrete approaches (see the example illustrated in Table 3.1).

Binary frequency consists of counting the number of initial representations of the information (objects) in intersection with the tile (an object can cover all or a part of several tiles). Thus, each representation in intersection with a tile increments the frequency of the latter by 1. Proportional frequency is based on the ratio of overlapping between an object and a tile: the frequency is thus incremented by a value between 0 and 1. Table 3.2 details the formulas dedicated to the calculation of these two types of frequencies.

Binary frequency	$freq(T_i) = \sum_{j=1}^{p} freq(T_i, O_j)$
Proportional frequency	$freqP(T_i) = \sum_{j=1}^{p} freq(T_i, O_j) * \frac{Surf(T_i,O_j)}{Surf(T_i)} * \frac{1}{NbTiles(O_j)}$

Table 3.2. *Formulas for calculating the frequency of a tile T_i - p: the number of objects in the initial index, $freq(T_i, O_j)$: frequency of the object O_j in the tile T_i (intersection), $Surf(T_i, O_j)$: surface of the object O_j in the tile T_i, $Surf(T_i)$: surface of the tile T_i, $NbTiles(O_j)$: number of tiles in intersection with the object O_j – taken from [PAL 10d]*

We then use these frequencies for calculating the weight associated with each tile invoked in a given document unit. We use four formulas of weighting. TF, TF·IDF and OkapiBM25 [MAN 08b] are applied to the weighting of tiles from binary frequency calculations. TF_p, an adaptation of TF, is applied to the weighting of tiles from proportional frequency calculations.

Table 3.3 presents these different formulas. The normalization of the frequencies calculated for TF and TF_p must be noted: thus, the weight of a tile in a document unit is divided by the total number of tiles invoked in this same unit.

These standardized indexes contain, for each tile, a list of tuples mainly composed of their weight (TF, TF.IDF, OkapiBM25 and TF_p), the identifier of the corresponding document and paragraph (document unit).

It is now possible to apply the IR models allowing the use of such generalized indexes composed of tiles.

3.4.2. *Spatial and temporal IR applied to tiling: PIV²*

The vectorial model of Salton [SAL 71, SAL 75], well-tried in IR, gives good results [BAE 99]. We apply it to spatial and temporal tiles: this consists of representing a set of tiles describing a document in the form of a first vector as well as the set of tiles corresponding to a query in the form of a second vector, then comparing these two vectors. The document repository is thus described by a matrix, as shown in Table 3.4 (D corresponds to a document, T to a tile and w_{ij} to the weight of the tile j for the document i). The score of a retrieved document is, for example, determined by calculating the scalar product [BAE 99, GOK 09].

Tile frequency (TF)	$W(t, Du) = TF(t, Du) = \frac{freq(t,Du)}{\sum_{i=1}^{n} freq(t_i)}$
TF·IDF	$W(t, Du) = TF(t, Du) * IDF(t) \quad with\ IDF(t) = log\left(\frac{NDu}{NDu_t}\right)$
OkapiBM25	$W(t, Du) = \left(\frac{(k_1+1)*TF(t,Du)}{(K+TF(t,Du))}\right) \quad with\ K = k_1 * [(1-b) + \frac{b*n}{advl}]$
TFp	$W(t, Du) = TFp(t, Du) = \frac{freqP(t,Du)}{\sum_{i=1}^{n} freq(t_i)}$
$freq(t, Du)$: frequency of the tile t in the document unit Du $freqP(t, Du)$: continuous frequency of the tile t in the document unit Du n: number of tiles in the document unit Du $\sum_{i=1}^{n} freq(t_i)$: cumulated number of occurrences of tiles in the document unit Du NDu_t: number of document units related to the tile t NDu: number of document units, k_1 = 1.2 b = 0.75, $advl$ = 900	

Table 3.3. *Weighting formulas applied to the standardized indexes, for a tile t and a document unit Du – taken from [PAL 10d]*

$$\begin{array}{c} \\ D_1 \\ D_2 \\ \vdots \\ D_n \end{array} \begin{array}{c} \begin{array}{cccc} T_1 & T_2 & \dots & T_t \end{array} \\ \begin{pmatrix} w_{11} & w_{21} & \dots & w_{t1} \\ w_{21} & w_{22} & \dots & w_{t2} \\ \vdots & \vdots & & \vdots \\ w_{n1} & w_{n2} & \dots & w_{tn} \end{pmatrix} \end{array}$$

Table 3.4. *Vectorial model: document-tile matrix*

Given that the information can be represented via different levels of generalization, the proposed multi-level tiling allows us to use the index of tiles most adapted to the range of the user's query.

Several tests described in [PAL 10a] will be discussed in section 3.5. These will mainly allow us to verify that the loss of precision due to tiling does not degrade the results of our new spatial ($PIV^2_{spatial}$ prototype) and temporal ($PIV^2_{temporal}$ prototype) IR approaches. These will also allow us to determine which tiling and which weighting formula give the best results for the spatial dimension and the temporal dimension.

We now have homogeneous spatial and temporal IRSs, because they are based on the same model of data representation and on the same IR model. In the following chapter, we will focus on the design and implementation of a GIRS combining the spatial, temporal and thematic dimensions.

3.4.3. *Multicriteria IR applied to tiling: PIV^3*

In digital libraries, IR presents a number of specificities. The content of the corpora generally references specific fields (fauna, flora, art, sport, etc.). Consequently, the needs as well as the vocabulary of the users are more precise. Moreover, the users of these repositories are familiar with the manipulation of the search operators. Thus, an advanced user interested in documents dealing with "Pyrénées mountains but not those of Gavarnie, in the 19th Century, if possible, unrelated to ascents" [PAL 10a] has to be able to express this type of need and navigate in the set of document fragments retrieved by the IR system. The more expressive the query is, the higher the number of associated criteria and operators will be. This must finally be completed by the particularity of geographic information that we approach via the spatial, temporal and thematic dimensions. We locate these propositions in the line of the study of Ray R. Larson [LAR 09] who shows the efficiency of GIR methods in the access to digital libraries.

During the process of need formulation, this category of users introduces more expressiveness with the notion of role such as "should", "has to", "must not" to give importance or priority to a criterion or to use a criterion as an exclusion key. The usual search engines (such as Google, Terrier or Lucene) propose two ways of formulating needs:

– *regular:* the user gives one or more keywords, the system combines the different sets of results without the intervention of the user;

– *advanced:* the user gives several keywords and specifies those which are mandatory or to be excluded and can assign a weight to others.

GIRSs propose different types of formulation similar to the regular type mentioned above. The user does not intervene in the aggregation of the results; the expressiveness of the queries is relatively weak. Consequently, we propose an IR approach which is:

– *multicriteria:* the query contains several search criteria to match documents;

– *geographic:* each criterion concerns the spatial, temporal or thematic dimension of the document repository;

– *expressive:* each criterion can be associated with a role that assigns a level of requirement or preference to it.

3.4.3.1. *Multicriteria IR: example of user-guided combination*

Let us resume the example of the user interested in the Pyrénées mountains and looking for past descriptions of these mountains. He or she would nevertheless like to exclude the documents dealing with the commune of Gavarnie and, as much as possible, avoid those centered around ascents.

Figure 3.4 corresponds to what could be a less ambiguous representation of this query. It contains five search criteria. Each corresponds to a particular geographic dimension. Finally, a predefined role can be associated with it.

Criterion	Pyrenees	Gavarnie	1800-1900	mountain	ascent
Dimension	Spatial	Spatial	Temporal	Thematic	Thematic
Role	Prerequisite	Exclusion	Promotion	Target	Demotion

Figure 3.4. *An example of representation of a multicriteria geographic query*

Figure 3.5 illustrates the five roles we propose potentially associating with a search criterion. For example, the ("must not") *Exclusion* is a role that excludes from the result list any document consistent with the criteria; the ("should not") *Demotion* pushes back in the result list any document consistent with the criteria; the ("should") *Promotion* promotes in the result list any document consistent with the criteria; the ("must") *Prerequisite* selects any document consistent with the criteria and, consequently, excludes from the result list any document inconsistent with the criteria; the ("should and must") *Target* selects and promotes any document consistent with the criteria and, consequently, excludes from the result list any document inconsistent with the criteria. Our proposition is original in the sense that the rejection and acceptance rules associated with a search criterion are thus reduced to five predefined roles.

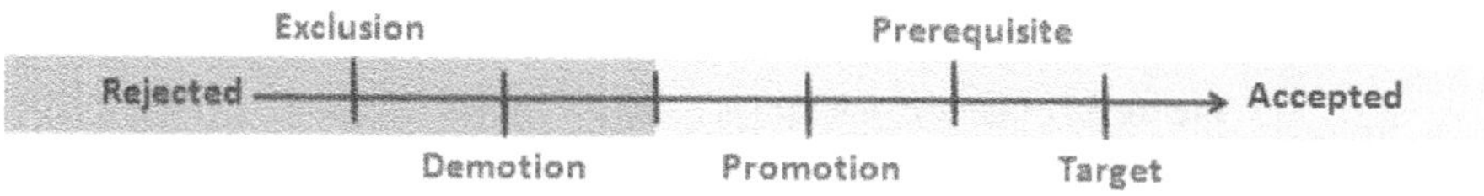

Figure 3.5. *Five predefined roles to be associated with the search criteria for a higher expressiveness*

Thus, according to their level of expertise, the users formulate their need in two different ways. First, regular-level users assign these different roles to the search criteria. By default, the *Promotion* role is in effect when no role is specified. Second, advanced-level users assign modal operators to the search criteria in order to modulate their range in a finer way. We propose the requirement operator of + and – type and the preference operator of ^ type. The ^ operator gradually penalizes or boosts a criterion by a weight coefficient ranging between $[-1, 1]$. Figure 3.6 illustrates the approach of an advanced user. The modal operators of requirement and preference specify the role associated with each criterion, independently of the dimension involved. The matching operators (intersection (∩), equality (=), etc.), on the other hand, are dependent on the spatial, temporal or thematic dimension of the criteria.

Criterion	Pyrenees	Gavarnie	1800-1900	mountain	ascent
Dimension					
Geography	Spatial	Spatial	Temporal	Thematic	Thematic
Operator	∩	=	∩	=	=
Role					
Requirement	+	-		+	
Preference	0	0	0.8	0.7	-0.5

Figure 3.6. *An example of advanced representation of a multicriteria geographic query – illustration taken from [PAL 10a]*

The correspondence between the five previously defined roles and the modal operators of requirement and preference is shown in Figure 3.7. To each of the five roles corresponds a combination of levels of requirement and preference defined by default.

Role	**Requirement**	**Preference**
Exclusion	-	
Demotion		-0.5
Promotion		+0.5
Prerequisite	+	
Target	+	+0.5

Figure 3.7. *Detailed description of the predefined roles and corresponding default values*

Figures 3.4 and 3.6 illustrate two levels of formalization of the multicriteria query which we have taken as an example. Let us note that, in this example, the temporal criterion “1800–1900” is not required but promoted as soon as it is satisfied in a document. The thematic criterion “mountain”, on the other hand, is at the same time required and promoted; a *Target* role is associated with it.

It must be noted that, in an advanced querying scenario (Figure 3.6), the adjustment of the requirements and preferences is performed directly by the user. However, certain combinations are incoherent (e.g. a $-$ requirement and a $+$ preference). It is therefore necessary to perform controls.

3.4.3.2. *Multicriteria IR: formalization of the partially compensatory CMRP approach*

We thus propose a new approach, inspired from those used in multicriteria decision support [MAR 99], in order to develop a more expressive IR. The aim is to enable the user to specify, for each criterion of their query (Figure 3.6), the corresponding geographic matching operator as well as the modal operators of requirement and of preference.

The partially compensatory aggregation model of query results which we propose can thus be modeled as a (C, M, R, P) quadruplet:

– criteria C $= (c_1, \ldots, c_n)$, where c_i is a search criterion of the query;

– matching operators M $= (m_1, \ldots, m_n)$, with $m_i : C \times D \rightarrow [0;1]$, and m_i refers to an external matching function used to compare any document $d \in D$ with the criterion c_i. The score assigned to the document for this criterion corresponds to a value between 0 and 1. Various operators may be available such as intersection, equality, inclusion and proximity. These correspond to the functions supported by the spatial, temporal and thematic IRSs, respectively;

– requirements R $= (r_1, \ldots, r_n)$, where r_i is a modal operator of requirement applied to the criterion c_i. It indicates the mandatory $(+)$, neutral (N) or exclusive $(-)$ character of this criterion;

– preferences P $= (p_1, \ldots, p_n)$, *with* $p_i \in \mathbb{R}$, and p_i weighs the search criterion c_i according to the corresponding level of importance desired by the user.

Thus, the CMRP model [PAL 12a] allows users to adjust the aggregation of results according to their expectations. The final score of a document is established by the partially compensatory aggregation function *RSV* described in equation [3.2]. This function weighs each intermediate score associated with a document for a given criterion according to the level of preference initially assigned to the criterion by the user. Thus, the score of a document d corresponds, for a set of n criteria, to the sum of the products of the preference level p_i by the score obtained for each criterion c_i (the m_i matching function applied to the document d with c_i as criterion). This sum is then normalized by the sum of the positive preference levels. Let us note that, if a mandatory criterion is not satisfied or, on the contrary, an exclusion criterion is

($d \nsim R$), it is not necessary to calculate the score because the document does not have to be retrieved ($RSV(d) = 0$).

$$RSV(d) = \begin{cases} 0 & \text{if } d \nsim R \\ \frac{1}{\sum_{i=1}^{n} p_i | p_i > 0} \cdot \sum_{i=1}^{n} p_i \cdot m_i(d, c_i) & \text{else} \end{cases} \quad [3.2]$$

In summary, we have proposed a new multicriteria IR model that extends those implemented in criterion aggregation, illustrated in [FAR 08]. CMRP presents several advantages:

– for each criterion, matching functions ($m_i(d, c_i)$) depend on dimension-dedicated IRSs;

– for each criterion, expressiveness (i.e. requirement or preference (p_i)) is independent of the invoked IRS;

– for each query, the aggregation of the results is partially compensatory: the $RSV(d)$ score computation takes into account the requirements and the levels of preference associated with each criterion by the user.

We present hereafter the PIV^3 platform that supports the implementation of different aggregation models such as CombMNZ [FOX 93], PSM [DA 09, DA 12] and CMRP [PAL 10a].

3.4.3.3. *Multicriteria IR: implementation on the PIV^3 platform*

The multicriteria CMRP IR model, described in the previous section, can be implemented on a single search engine or several search engines federated by a meta-engine. As recommended by Rasolofo *et al.* [RAS 03], we have adopted the second approach in order to develop the PIV^3 multicriteria IR platform.

Figure 3.8 shown a meta-search engine as a broker that splits a query into sub-queries dedicated to the targeted search engines. This meta-engine also supports a result lists aggregation process that produces a single final list.

Thus, the PIV^3 meta-engine, detailed in [PAL 10a], implements the CMRP model following the global architecture presented in Figure 3.8. Three "drivers" feature the PIV^2_spatial, PIV^2_temporal and PIV^2_thematic (based on the Terrier [OUN 05] IR engine) search engines. Spatial, temporal and term-based specific matching functions are declared in these drivers. PIV^3 thus federates three mono-dimensional IRSs. It breaks down a multicriteria query into sub-queries and relays each of them to the corresponding IRS. PIV^3 supports different result aggregation models, including the CMRP model that, as we have seen, allows us to extend the expressiveness associated with each criterion. In this case, PIV^3 performs the partially compensatory aggregation formula *RSV* (equation [3.2] of the previous section) in order to take into account the roles associated with each search criterion and merge the results in a single final list. Let us note that each mono-dimensional IRS relies on a tile-based generalized representation of information. Moreover, each IRS implements the vectorial IR model.

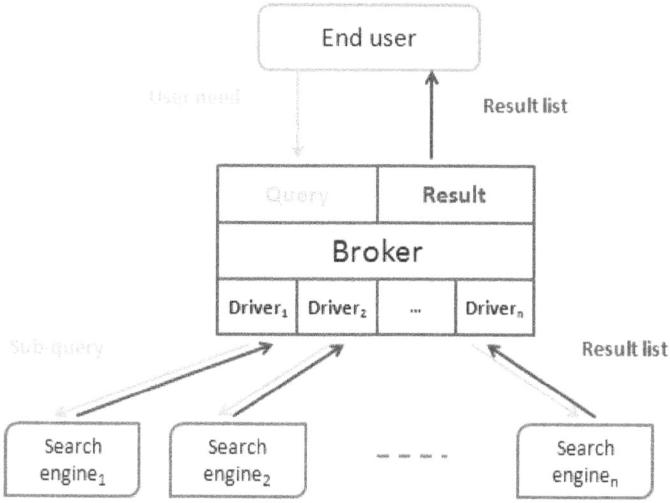

Figure 3.8. *A meta-search engine according to Rasolofo et al. [RAS 03]*

More generally, the PIV^3 meta-engine can also federate any type of search engine and implement different result aggregation models.

In the following section, we evaluate the CMRP model implemented in the PIV^3 meta-engine.

3.5. Evaluation and discussion

As we have emphasized in Figure 2.4 in section 2.3.7, there is, to our knowledge, no GIRS evaluation framework combining the spatial, temporal and thematic dimensions of information. Thus, TREC [VOO 05] is a reference campaign in IR that allows us to evaluate IRSs with respect to the thematic dimension. There are few studies concerning the evaluation of these two other dimensions of geographic information. The temporal dimension has been the object of the TEMPEVAL evaluation framework [VER 09]. Furthermore, Bucher *et al.* [BUC 05] have proposed the simultaneous consideration of two dimensions: spatial and thematic. This proposition can be found in the GEOCLEF task [GEY 05] of the CLEF framework [PET 01]. CLEF has mainly enabled the evaluation of classic thematic IRSs in IR such as Lemur [OGI 01], Lucene [GOS 05] and

Terrier [OUN 05], as reported in [PER 08]. The existing GIR works (presented in Table 2.1 of Chapter 2) have also been evaluated from the point of view of the size of the indexes, the time to construct the indexes and the query times [VAI 05]. These evaluations would only benefit from being put into perspective with other measures such as the precision or the recall of the considered GIRSs. It is therefore impossible to compare the search engines driven to process simultaneously the three geographic dimensions. Therefore, in order to respond to this need, we propose an experimental framework dedicated to GIRSs. This experimental framework is focused on emphasizing the existing know-how of campaigns, such as TREC and GEOCLEF, while integrating the specificities relative to geographic information.

We then implement this framework for the evaluation of our IR prototypes PIV^2 and PIV^3 based on the information indexing and retrieval by tiling.

3.5.1. *Evaluation framework of geographic IRSs: proposal for a test collection and an experimental protocol*

In addition to the description of evaluation campaigns presented in Table 2.1 in section 2.3.6, we recall that a test collection contains the following:

1) A set of n "topics" representing users' information needs. Each topic is at least characterized by a title (a keyword-based query), a description (usually a sentence in natural language) and a narrative (a detailed explanation of expected information as well as criteria for judging a document as relevant or non-relevant). Buckley and Voorhees [BUC 00] show that at least 25 *topics* are necessary to perform statistically significant analyzes. Let us note, however, that the standard of TREC is equal to 50 *topics*.

2) The "corpus" of documents, some of which are relevant for the proposed *topics*. A regular TREC corpus for a classic ad hoc task is made up of 800,000 documents and more [VOO 05].

3) The "qrels", TREC term denoting the *query relevance judgments*, associating each *topic* with the set of relevant documents. Because the corpus is too large to be analyzed exhaustively, IR evaluation frameworks rely on the *pooling* technique. Thus, for each *topic*, a *pool* of documents is created from the top 100 documents retrieved by the participants' IRSs, duplicates being removed. The hypothesis is that the number and the diversity of the IRSs contributing to the *pool* will allow us to find most of the relevant documents. Finally, a human *assessor* examines each document of the *pool* in order to evaluate whether or not it matches the information need specified in the considered *topic*. The document is then qualified as relevant or non-relevant.

Such test collections have been developed several times in evaluation frameworks such as TREC and GEOCLEF. Let us note that these do not take into account the three dimensions of geographic information. This is the reason why we propose the creation of a collection including the following:

1) *Topics* covering all or a part of the three dimensions. For example, a *topic* could be entitled "Transhumance in the Pyrénées in the 19th Century" with the narrative "Will be considered relevant the documents invoking transhumance or related events (daily life of a shepherd in the pasture) in the Pyrénées mountains between the years 1800 and 1899".

2) A *corpus* dealing with the three dimensions: the conventionally considered thematic aspect is completed by spatial and temporal elements.

3) *Qrels* by dimension for which the assessor evaluates the consistency between each of the three dimensions (thematic, spatial and temporal) and the document. The mere presence of the three dimensions in the document is not sufficient to deduce its relevancy for the query. Let us consider, for example, the case of a document dealing with the spa industry, then mentioning "Gavarnie" as the birth place of the narrator. Although spatially relevant, it does not match the query "spa industry in Gavarnie". Such a subtlety requires the assessment of the global match between the query and the document. In judging a given document, the assessor first evaluates its adequacy according to each of the three dimensions. Not to overwhelm assessors, we opted for a per dimension binary judgment: a document is either relevant or non-relevant to the considered query and dimension. This rationale is akin to the conclusions of Bucher *et al.* [BUC 05] who underline the fact that the dimensionally gradual assessments are "unnecessarily cumbersome". Finally, considering the three per-dimension binary judgments, as well as the aforementioned global binary judgment, we compute the document relevance value $v \in \{0; 1; 2; 3; 4\}$. This both represents the number of satisfied dimensions (from 1 to 3) and global relevance (4). No assumption was made regarding the relative importance of dimensions: they were equitably considered.

4) *Geographic resources* necessary, on the one hand, for the geo-referencing of the SFs and the interpretation of TFs contained in the corpus, on the other.

It is on this basis that, in the following section, we describe the test collections and experimental protocols implemented to evaluate the PIV^2 and PIV^3 prototypes based on the approach of indexing and IR by tiling.

3.5.2. *Evaluation of the spatial and temporal IR applied to tiling*

To validate our different propositions, we have setup several experiments. We have evaluated the interest of our standardization approach mainly concerning the loss of precision induced by tiling. For this, we have compared our spatial PIV IRS with the PIV^2 IRS integrating the process of spatial standardization. Following the same principle, we have compared the temporal PIV IRS with the PIV^2 IRS integrating the process of temporal standardization. A detailed description of these experiments is presented in [PAL 09], [PAL 10a] and [PAL 10d], as well as in [SAL 12]. These studies answer the following questions:

– Does the approach of standardization have an interest? Does it induce a too significant loss of precision capable of very strongly degrading the quality of results retrieved by a spatial or temporal IRS?

– Does a tiling range or a weighting formula differentiate itself in the particular context of our corpus?

– Does a different weighting formula have to be used for the type of relationship (spatial/temporal adjacency, inclusion, orientation, etc.) processed?

– Does the use of an index adapted to the spatial or temporal scale of the query improve the results?

3.5.2.1. *Evaluation of the spatial IR*

This analysis consists of evaluating the interest of spatial standardization. For this, we have compared our standardized spatial PIV^2 IRS with the initial spatial PIV IRS: 40 topics containing ASFs and RSFs of different ranges were tested.

Prototype	*Index*	*IR model*	*Improvement*
spatial PIV	geometry	ad-hoc	base line
spatial PIV^2	tile	vectorial	+13%

Figure 3.9. *Comparison of the PIV and PIV^2 spatial IRSs*

As we can see in Figure 3.9, PIV^2 gives results superior in precision to those of PIV by 13%. The loss of precision induced by tiling is balanced by the use of frequency calculations.

3.5.2.2. *Evaluation of the temporal IR*

This analysis consists of evaluating the temporal standardization. As for the spatial information, we have compared our PIV^2 standardized temporal IR system with the initial PIV temporal IR system: 35 topics containing ATFs and RTFs of different ranges have been tested.

A summary of the results presented in Figure 3.10 shows that the temporal standardization associated with PIV^2 gives similar results to those of PIV (–0.5%). Contrary to spatial, temporal information is less in number and paragraphs containing several temporal references are rare. This indeed explains that the use of tile frequency for the results score (RSV) computation does not contribute to the improvement of their precision.

Prototype	*Index*	*IR model*	*Improvement*
temporal PIV	Interval	ad-hoc	base line
temporal PIV^2	tile	vectorial	-0.5%

Figure 3.10. *Comparison of the PIV and PIV^2 temporal IRSs*

3.5.2.3. *Summary of the evaluation of PIV^2*

Spatial standardization coupled with the usage of classic IR formulas in PIV^2 improves the results of the spatial PIV IR approach. Temporal standardization coupled with the usage of classic IR formulas in PIV^2 has not shown significant changes with respect to the temporal PIV IR approach. To answer the three other questions, the experiments described in [PAL 10a] have shown that:

– communal tiling (respectively monthly) coupled with the TF_p weighting formula gives the best results;

– no matter the spatial (respectively temporal) relationship involved in the information retrieval, the TF_p weighting formula gives the best results;

– the use of an index adapted to the spatial or temporal scale of the query improves the results.

3.5.3. *Evaluation of the multicriteria IR applied to tiling*

We have compiled a test collection MIDR_2010[3] containing a *corpus*, *topics*, the corresponding *qrels* and *geographic resources* [PAL 10c]. We have based our study on this repository to evaluate the multicriteria IR according to the CombMNZ model (for a totally compensatory aggregation), on the one hand, and according to the CMRP model (for a partially compensatory aggregation), on the other.

3.5.3.1. *Evaluation of the totally compensatory multicriteria IR*

This experiment has been carried out on the MIDR_2010 test collection for 44 *topics* containing spatial, temporal and/or thematic criteria. We have evaluated the PIV^3 GIRS implementing the CombMNZ totally compensatory aggregation model:

3 This is a French test collection accessible at http://t2i.univ-pau.fr/MIDR.

for a given document, the sum of the scores obtained for the different search criteria is weighed by the number of IRSs having retrieved this document [PAL 10a, SAL 12].

Thus, PIV^3, configured according to the CombMNZ model, improves the precision of the Terrier, PIV^2_spatial or PIV^2_temporal IRS, alone [PAL 10a, SAL 12]. As shown in Figure 3.11, the improvement with respect to the bag of words-type (Terrier) thematic IR is equal to 66%. Let us note that the spatial, temporal and thematic criteria are of the same importance here. Moreover, the detailed analysis of the results shows a small number of documents returned simultaneously for the three dimensions, consequently showing that the three dimensions are complementary. A detailed description of these experiments is presented in [PAL 10a, PAL 10c, CAB 11] and [SAL 12].

Prototype	*Index*	*Weighting method*	*Improvement*
Terrier	term	TF.IDF	base line
Terrier	term	OkapiBM25	0%
PIV^3 CombMNZ	tile	compensatory aggregation	+66%

Figure 3.11. *Comparison of the* Terrier *and PIV^3 CombMNZ IRSs*

3.5.3.2. *Evaluation of the partially compensatory multicriteria IR*

No test collection exists which covers all the three geographic dimensions and which, in addition, proposes *topics* associating requirements or levels of preference with the search criteria. So, this experiment has been carried out on the MIDR_2010 test collection for 10 expressive geographic *topics* targeting territories, periods and subjects invoked in this corpus (e.g. "Pyrénées mountains but not those of Gavarnie, in the 19th Century, if possible, unrelated to ascents"). We have experimented the PIV^3 GIRS endowed with the CombMNZ aggregation models, on the one hand, and CMRP, on the other. In this latter case, the different scores of a document are affected by coefficients, which are the levels of preference expressed in the query.

The PIV^3 CMRP GIRS, based on a partially compensatory aggregation, improves the precision of the PIV^3 CombMNZ GIRS, based on a totally compensatory aggregation. As shown in Figure 3.12, this improvement is equal to 54%. Let us note that here the set of queries associates levels of preference and requirements with the criteria. The observed difference illustrates well the fact that CMRP allows users to refine the roles for each criterion, contrary to CombMNZ. Indeed, CombMNZ supports a totally compensatory aggregation while CMRP offers an increased power of expression to the users. A detailed description of this experiment is presented in [PAL 12a] and [SAL 12]. We have to continue our tests and evaluate more than 25 *topics* in order to confirm this trend.

Prototype	Index	Weighting method	Improvement
PIV³ CombMNZ	tile	compensatory aggregation	base line
PIV³ COPE	tile	partially compensatory aggregation	+54%

Figure 3.12. *Comparison of the PIV³ CombMNZ and PIV³ CMRP IRSs*

3.5.3.3. *Summary of the evaluation of PIV³*

The testing of PIV³ endowed with the CombMNZ model, applied to the MIDR_2010 test collection for the multicriteria GIR, validates our starting hypothesis: "a totally compensatory aggregation of results (CombMNZ) from spatial, temporal and thematic IRSs gives better results than a single thematic IRS or two-by-two pairings of spatial, temporal and thematic IRSs" [PAL 12a].

The testing of PIV³ endowed with the CMRP model, applied to the MIDR_2010 test collection for the multicriteria GIR, validates our second starting hypothesis: "a partially compensatory aggregation of results from spatial, temporal and thematic IRSs, offering more expressive power to the user, gives better results than a totally compensatory aggregation" [PAL 12a]. The CMRP model allows the user to associate specific roles with each search criterion, which explains the higher quality results [PAL 12a].

In its current state, the proposed experiment presents at least two limits. First, with its 5,645 paragraphs totaling 3.7 MB, the MIDR_2010 test collection is of a very small volume in comparison to the TREC collections. Second, we have performed analyzes from 44 then from 10 *topics*. We continue the efforts of manual assessment of the documents allowing us to obtain more *topics* to analyze.

Furthermore, we have also analyzed the contribution of expressive queries integrating modal operators of requirement and preference in a single thematic context. We have mainly studied the TREC-8 [VOO 99] collection to measure the potential of improving the precision brought by the use of operators. The use of different operators has been studied in [SIL 99], [SPI 01], [EAS 03], [WHI 07] and [AUL 10] by analyzing the *logs* of such search engines as Altavista, Excite, Google, MSN Search and Yahoo!. Globally, these studies have shown that only 20% of all queries contain operators and that the use of operators improves relatively little the quality of the results. Thus, independently of the users' profile, we have studied the expected potential of improvement in a context of optimal usage of operators [PAL 10b, HUB 11]. We have thus shown that the expressive *topics* prevail over the standard *topics* in 68% of the cases. When the terms of an expressive *topic* are well chosen, there is always a multicriteria query that could improve the results of the search [PAL 10c].

Finally, this final experiment has also allowed us to show that it is thus possible to federate search engines which do not support the modal operators of requirement and preference.

3.6. Summary

This chapter is devoted to GIR. We have designed, implemented and tested a number of propositions contributing to the development of a multicriteria IRS, adapted to the spatial, temporal and thematic dimensions of geographic information.

We have thus proposed a generic approach to standardization applicable to each of the dimensions of the indexed geographic information [PAL 12b]. The $\text{PIV}^2_{\text{spatial}}$ and $\text{PIV}^2_{\text{temporal}}$ prototypes implement spatial and temporal indexes generalized by administrative and calendar tilings, respectively.

Moreover, we have developed a partially compensatory multicriteria IR approach (CMRP model) supporting roles which, associated with every search criterion, increase its expressive power [PAL 12a]. The PIV^3 prototype supports the multicriteria IR and combines results from spatial, temporal and thematic IRSs: in the framework of our tests, PIV^3 brings together $\text{PIV}^2_{\text{spatial}}$, $\text{PIV}^2_{\text{temporal}}$ and Terrier [OUN 05].

3.6.1. *Contributions*

We have studied information needs formulated in the contexts of IR linked to digital libraries and expressive queries. We propose a partially compensatory multicriteria query model in which each criterion can be refined by a role describing its importance in the query. Thus, the CMRP model supports the definition and the processing of preferences and requirements in such queries. The requirements are expressed by modal operators that emphasize the mandatory, exclusive or neutral character of a criterion, whereas the preferences are expressed over a criterion by other operators such as factors of promotion and demotion. The CMRP model integrates these operators in a partially compensatory aggregation function combining the lists of results from each of the search criteria. These modal operators satisfy a higher need for expressiveness and allow the promotion or demotion of documents in the final list of results. We have designed the PIV^3 GIR engine (see Figure 3.13) that implements CMRP for purposes of validation and testing of our propositions. This system takes into account the geographic dimensions of information by involving domain-specific IRSs. Each domain IRS being dedicated, respectively, to spatial, temporal or thematic information, the search engine allows the end user to exploit these dimensions in their query. Each search criterion is therefore part of a particular dimension and can be associated with roles.

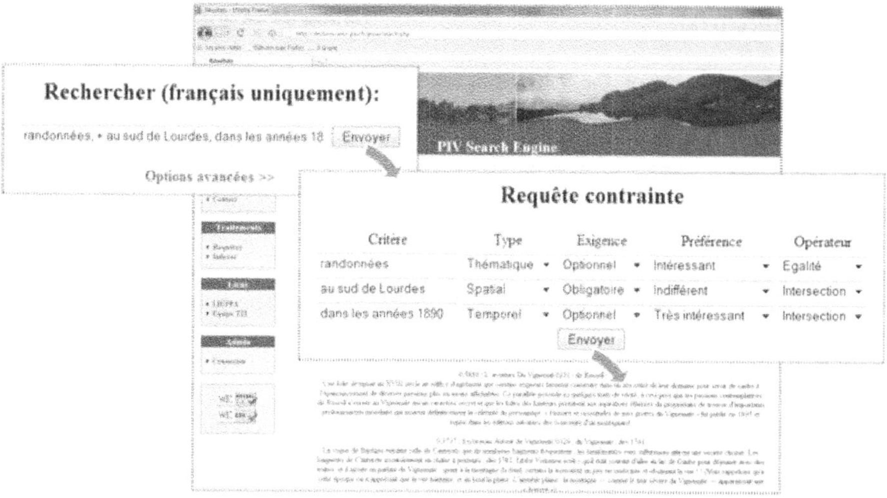

Figure 3.13. *Illustration of the multicriteria IR supported by PIV*^{3} *prototype – "hikes, (+) to the south of Lourdes, (^0.8) in the 1890s"*

These studies have allowed us to develop different prototype versions mainly contributing to the CIDRI project, described in Chapter 1 of this book:

– a model of generalization of indexed data representation for multicriteria IR;

– a model of multicriteria IR for taking into account geographic information needs with a strong power of expression;

– corresponding process flows for the implementation and testing of PIV^2 and PIV^3 IRSs.

An initial experiment based on geographic queries and the MIDR_2010 test collection shows that, in terms of precision, the PIV^3 GIRS associated with the CMRP model surpasses the configuration integrating the CombMNZ [FOX 93] model by +54.16%. A final experiment based only on thematic queries of TREC-8 shows that PIV^3 associated with the CMRP model improves the precision of search results in a significant way (+26.96%) in a context that implements expressive *topics*.

These studies are based on a first set of results presented in Chapter 2 and are the fruit of numerous collaborations having mainly involved Guillaume Cabanac and Gilles Hubert, academic researchers at the IRIT, and Damien Palacio, a doctoral student at the LIUPPA.

3.6.2. *Perspectives*

In the framework of the CMRP model and its implementation in PIV^3, the number and power of expression of the IR operators correspond to the proposition of a dynamically customizable formula of result aggregation. This approach makes the task of the end users more difficult concerning the elaboration of queries as well as the control of their range. We could study two categories of research interfaces. The first category, dedicated to regular users, would enable them to choose between predefined roles in order to add expressiveness to the search criteria: see Figure 3.13 with customizable requirement and preference criteria. The second, dedicated to advanced users, would allow them to customize their queries by specifying modal ($\hat{}$, $+$, $-$) and matching operators ($=$, $\subseteq$, $\cap$).

Concerning the MIDR_2010 test collection, a greater corpus and set of *topics* have to be established in order to conduct a larger scale experiment. However, this evaluation provides several interesting first results and can simulate the use of unexploited operators in GIR and IR systems (e.g. Lemur [OGI 01] or Lucene [GOS 05]). Furthermore, we can note that the completeness of the proposed language is not assured, even though CMRP supports the "and" and "or" operators through those of requirement and preference.

Finally, let us note that while the principle of multicriteria IR is well suited to geographic information, as we have shown by conducting an experiment on the TREC-8 collection, it could equally be applied to the combination of different textual IRSs and/or other specialized ones such as opinion IRSs [ROC 09, LI 10], for example.

4

General Conclusion

4.1. Summary

4.1.1. *Contributions to the access by geographic content to textual corpora*

The propositions discussed in the previous chapters target the spatial, temporal and thematic dimensions in the process of information retrieval (IR) in general and, more particularly, the access to textual corpora in which the spatial and temporal dimensions are strongly represented.

In the works of published literature relative to the geographic IR, we have mainly taken interest in the following challenges:

1) the recognition and resolution of spatial and temporal named features;

2) the spatial and temporal indexing for purposes of IR;

3) the matching of document/query couples and the calculation of relevance scores devoted to spatial IR on the one hand and temporal IR on the other hand;

4) the multicriteria IR combining the spatial, temporal and thematic dimensions;

5) the evaluation of such IR systems.

The objective of T2I team of the laboratory of LIUPPA is the *"Processing of spatial, temporal and thematic information for the adaptation of interaction to the context and the user"*. The extraction and interpretation of spatial and temporal information contained within textual documents is the center of these concerns. The contribution of the team thus concerns the processes aiming at the recognition and semantic analysis of such information in texts in order to build the corresponding annotations. It also concerns the proposition of symbolic spatial and temporal data representation models necessary for this recognition, interpretation and representation not only of toponymic or calendar references but also of spatial or temporal relations. Thus, the propositions relative to challenge no. 1 presented above correspond to the work conducted under the direction of Gaio [GAI 08].

These results of extraction and annotation were then used for the purposes of multicriteria information indexing and retrieval taking into account the spatial, temporal and thematic characteristics contained within the documents and the queries. These propositions, relative to challenges 2, 3, 4 and 5, also correspond to the work of the T2I team of the laboratory of LIUPPA conducted, in particular, in collaboration with researchers of the laboratory of IRIT [PAL 12a, PAL 10c].

With this in mind, the general issues for which we have proposed solutions were: (1) *What models of representation and retrieval of spatial and temporal information should be proposed for the access by geographic content to textual corpora?* (2) *What core model of representation and retrieval of information should be proposed in order to prepare for the combined use of these dimensions?* (3) *What advanced aggregation operators should be proposed and implemented for a multicriteria IR combining several information retrieval system (IRSs)?*

The first set of contributions focused on the spatial and temporal IR in textual documents in order to answer question (1). The second set of contributions dealt with the generalization of data representation and aggregation of results in the framework of IR in order to answer questions (2) and (3), respectively.

4.1.2. *Spatial and temporal IR in texts*

The work is divided into two main contributions.

The first contribution corresponds to the pieces of work that, as we have specified earlier, target annotation. And so, to repeat, we first of all proposed models of spatial and temporal information representation. Then, we designed a method of extraction and interpretation of spatial information, based on our model and on a specific semantic process. Similar procedures were performed for the implementation of a method of extraction and interpretation devoted to temporal information. These models and methods are applicable to non-structured textual corpora. In the context of the T2I team of the laboratory of LIUPPA, the spatial and temporal annotations resulting from these processes are used in the scenarios of IR or computer-assisted reading of documents, for example.

Thus, in the continuity of these propositions, the second contribution targets the indexing and retrieval of spatial and temporal information. We stored, in indexes, the results of our approach to approximating symbolic representations by numeric representations (geo-references, date intervals). We then developed a model of spatial IR. Finally, we developed, in a similar way, a model of temporal IR.

As illustrated in Figure 1.6 of Chapter 1, these propositions have been mainly implemented in the GEOSEM2, GEOTOPIA, GEOCIME and GEONTO projects, and have been tested in numerous versions of the PIV prototype. These experiments have enabled us to validate the following working hypotheses:

– A devoted spatial IRS gives better results than a classic thematic IRS for IR composed of only spatial criteria.

– A devoted temporal IRS gives better results than a classic thematic IRS for IR composed of only temporal criteria.

– A "rough" coupling of spatial, temporal and thematic IRSs gives better results than a classic thematic IRS for multicriteria IR despite the numerous possible biases.

4.1.3. *Multicriteria IR in texts*

These pieces of work have the objective of defining a meta-engine of geographic IR, which federates spatial, temporal and thematic IRSs. They are therefore based on the previous results relative to the spatial and temporal IR and are divided into two main contributions.

Our first contribution targets the generalization of data representations. To avoid possible biases, we have chosen to standardize the data representations and the approaches of processing data relative to the different dimensions. Thus, from the indexed representations of spatial and temporal information, we build higher level indexes, suitable for the implementation of IR models such as the vectorial model.

Our second contribution targets the partially compensatory aggregation of results in a context of spatial, temporal and thematic IRSs federation. The aggregation model we developed offers modal operators allowing us to associate a level of preference and requirement with each search criteria. This model integrates the matching operators supported by the federated IRSs and extends their expressiveness using modal operators.

As illustrated in Figure 1.6 in section 1.4.3, these propositions have been implemented in the CIDRI project, and have been tested in numerous versions of the PIV^2 and PIV^3 prototypes. These experiments have allowed us to validate the following working hypotheses:

– The generalization of data representation is adapted to the spatial and temporal information.

– The generalization associated with the vectorial model does not imply a loss of precision with respect to the initially proposed devoted IRSs.

– A "classic" coupling of spatial, temporal and thematic IRSs supported by the CombMNZ model gives better results than a single thematic IRS or a two-by-two pairing of spatial, temporal and thematic IRSs.

– An "advanced" coupling of spatial, temporal and thematic IRSs, supported by our partially compensatory aggregation model called CMRP, gives better results than a "classic" coupling. It offers, in addition, more power of expression to the user.

4.2. Perspectives

The central element of our concerns is thus the geographic information illustrated at the beginning of this book by Figure 1.2 in section 1.2.2. We consider geographic information as a molecule not only containing a spatial component, but also a temporal component and a thematic, or "phenomenon" component [USE 96, GAI 01].

Let us recall that our work's focus is accessing the content of non-structured geographic textual corpora. The experiments relative to the different versions of our prototypes, carried out on such corpora, have allowed us to observe information making and reference to places spatial indications, as well as descriptions of landscapes, temporal indicators and dates, implying a pronounced interest of these documents for the geographic aspect. We have mainly worked on travelogues (see an excerpt in Figure 1.1 in section 1.2.2) in which the authors frequently use an identical logical structure: the text is divided into paragraphs describing a part of their travel. Each part can consist of a description of an itinerary, a stage, a point of view, an observation and/or an event, etc. Although the retrieval and availability of information from travelogues remains a current issue [HAO 10], our results can, nevertheless, be applied to other types of corpora such as tourist guides, hiking guides as well as travel description blogs whose structure of discourse also presents a certain homogeneity.

Numerous approaches are focused on the *thematic* dimension in IR today, whereas others, also on the increase, associate the *spatial*, *temporal* and *thematic* dimensions in geographic information retrieval (GIR). Inspired by successive GIR conferences [PUR 11], here is a non-exhaustive list of open research problems in these fields:

1) Specification of data models taking into account the representation of relations between the spatial, temporal and thematic dimensions [PER 07, EZZ 11, MAT 11, HOF 12].

2) Emergence of domain ontologies for indexing documents, contextualizing queries or guiding matchings [LIN 06, BAZ 07, DUD 10, MEI 10, POO 10, DAO 11, FER 11, BAS 12, LAT 12, OHT 12].

3) Detection of geographic references in the form of place names or calendar features, and their corresponding qualifiers in natural language within texts [PUR 11, LEI 11].

4) Disambiguation of spatial features (SFs) [LEI 11].

5) Interpretation of the semantics carried by spatial, temporal and thematic references [MAL 00, PER 07, SCH 07, BAT 11, MAT 11]. For example, Leidner and Lieberman [LEI 11] as well as Purves and Jones [PUR 11] show interest in SFs such as "40 km to the north of Berlin", which describes a vague zone that has to be interpreted by taking into account the meaning carried by "40 km", "north" and the place of reference "Berlin".

6) Indexing of documents by taking into account the three geographic dimensions: for example, Vaid *et al.* [VAI 05] as well as Purves and Jones [PUR 11] show interest in the spatial and thematic dimensions.

7) Matching and ranking of relevant results based on one or more geographic dimensions [VAI 05, LIN 06, BUS 09b, URE 10, ALO 11, MAT 11, EGO 11, FER 11].

8) Development of user interfaces [PUR 11].

9) Conception of frameworks and resources devoted to the evaluation of geographic information retrieval system (GIRSs) [MAN 11, CAR 11, PUR 11].

In the T2I team of the laboratory of LIUPPA, we have chosen to work with a priority on space and time, which are generally easier to process entry points, because they are more stable in the language (e.g. recognition of named entities) and well formalized, when working on several dimensions of geographic information including the theme.

As part of this reflection, which consists of proposing perspectives, we propose extending these pieces of work following two main axes that we will qualify as being intra- and interdimensional.

Concerning the intradimensional axis, we propose reworking on our models of representation and reasoning. For the thematic dimension, we will review the theme, initially dealt with as a set of terms, and we propose a thematic tiling based on the definition of domain ontologies. For the spatial dimension, we are working on the approximation of the spatial information by more accurate representations than bounding boxes. This reflection targets challenges 2 and 5 mentioned earlier.

Concerning the interdimensional axis, we are working on a challenge identified [ALO 11] that, in temporal and thematic IR, raises the following question "should a relevant document in which the *textual match* and the *temporal match* are far away from each other be penalized?". It is mainly a question of identifying the relationships between spatial, temporal or thematic information detected in a text. This reflection targets challenge no. 7. Finally, we oppose an approach that we qualify as *qualitative* against the specialized IRS federating approaches. From knowledge bases, this *qualitative* approach consists of describing the spatial, temporal and thematic information with terms with the same meaning (toponyms, dates, events, etc.). It is a form of vocabulary extension through the use of specific gazetteers, which brings the three geographic dimensions down to this *qualitative* expansion. This reflection also targets challenge no. 7.

4.2.1. *Intradimensional axis*

We have developed a GIR system combining spatial, temporal and thematic IRSs. However, the thematic part is at the moment limited to the full-text Terrier IRS [OUN 05]. As with the spatial and temporal dimensions, we propose the generalization of thematic information of the first-level index created by the Terrier IRS. These perspectives are presented in section 4.2.1.1.

Concerning the spatial dimension, we have, in part, worked on representations of spatial information approximated by the bounding boxes. We envision the study of more accurate spatial representation modes in section 4.2.1.2.

4.2.1.1. *Thematic dimension*

We propose the use of a domain ontology: the concepts can be assimilated to the tiles generated for the spatial or temporal dimensions (e.g. administrative or calendar grids, respectively). Thus, the generalization of thematic information by tiling results in the creation of a second-level index, which is more synthetic and has a larger range than terms. It can be implemented by the following steps:

– Design of a thematic grid from a domain ontology: each concept corresponds to a tile, and each hierarchical level of the ontology corresponds to a level of abstraction (a botanic ontology composed of a classification of type *kingdom, class, order, family*, etc.).

– Analysis of text and extraction of meaningful terms with a thematic IRS (a set of lemmas extracted from the analyzed text: "gladiolus", "bloom", "summer", etc.).

– Projection of the terms onto the thematic grid and calculation of the tile frequencies (e.g. the "gladiolus" taxon of the botanic ontology retrieved among the extracted elements of the text).

As illustrated in [ZGH 08], the indexes created in such a way can be used by classic IR models such as the vectorial model. Moreover, in accordance with the propositions described in [LEG 10], these indexes can also be used by semantic IR models implementing similarity calculations between concepts such as the Wu and Palmer measures [WU 94], those of Resnik [RES 95], Lin [LIN 98] as well as Proxigenea [DUD 10].

In the framework of the ANR MOANO project, discussed in Chapter 1, we contribute to the construction of a botanic ontology and to the design of a system of indexing and retrieval of botanic information using this ontology. We reuse the ontological and terminological resource (OTR) model and the information indexing and retrieval engine TextViz developed in the framework of the ANR Dynamo[1] project [REY 09]. The current work focuses on the digitalization of "plant" files extracted from the Clause Villemorin guide[2] and their indexing based on a botanic ontology built in a semiautomatic way from resources taken from devoted Websites. Our contribution aims at information indexing and the retrieval in textual corpora. Following the previously mentioned ANR Dynamo project, we propose implementing an indexing Web service that, in input, takes a domain ontology and a text and, as output, produces the same text enriched by annotations corresponding to the concepts of the ontology. Concerning the IR, we have experimented queries focusing on flowers, their colors, seasonal properties, diseases, etc. The results of TextViz are sometimes particularly noisy (i.e. many non-relevant retrieved documents): Figure 4.1 presents a subset of results obtained for the query *Q7* ("Which flowers blossom in winter?"). The concepts of "flower" and "winter" have been well detected in the query and in each of the documents. However, there is no relation of blossoming between *winter* and the *plant* considered in the documents *agapanthe0.xml* and *buglosse1.xml*. Let us therefore consider the following approach in order to avoid this kind of side effect:

1) Use the concepts extracted from the query to look for relations that, in the ontology, link these concepts (e.g. *floweringPeriod, plantingPeriod*).

2) Select the relation expressed in the query among the propositions in this list (e.g. *floweringPeriod*).

3) Search all the flowers in the ontology that are associated with the target concept via this relation (e.g. "flower" in the relation *floweringPeriod* with "winter").

4) Finally, search for the documents of the corpus dealing with the flowers extracted from the ontology in such a way.

Q7: Which flowers blossom in winter?	
agapanthe0.xml	"flower" [...] "fears the harshness of *winter*"
chrysantheme7.xml	"flower" [...]"blossom until **winter**"
buglosse1.xml	"flower" [...]"protect from **winter** with a cover"

Figure 4.1. *Example of a query submitted to the TextViz IR engine*

1 http://www.irit.fr/dynamo/.

2 http://www.vilmorin-jardin.fr/vilmorin–graines-et-semences-pour-le-jardin-9-11-1-1.htm.

This is a way of filtering the set of documents resulting from the query presented in Figure 4.1 or even of completing it with new documents. This approach consists of using the relations described in an ontology in order to build the query result.

To our knowledge, only the CITER GIRS [PFO 09] uses themes coming from an ontology: it is only in the phase of IR that the user selects a theme in the proposed list for the search.

4.2.1.2. *Spatial dimension*

We are interested in the interpretation of spatial information detected in texts. Our objective is to make more accurate the approximation of the SFs indexed by the PIV process flow.

The current version of the PIV system proposes specific functions of calculating spatial relation representations [LES 07, SAL 08]. These functions are mainly based on propositions contributing to the interpretation of relations of adjacency and inclusion made by Malandain [MAL 00].

Figure 4.2 illustrates the *adjacency* function supporting the calculation of a polygon from an initial geometry. This function contains two parameters: the geometry *geom* of the reference SF and the distance *d* corresponding to the scope associated with this adjacency relation. It proposes a numeric interpretation of information expressed by phrases such as "next to, close to, around, in the periphery of, in the proximity of, in the surroundings of, nearby, encircling".

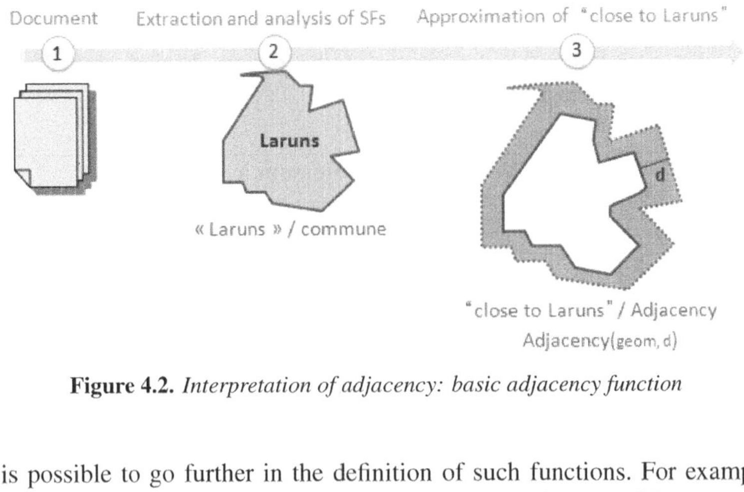

Figure 4.2. *Interpretation of adjacency: basic adjacency function*

It is possible to go further in the definition of such functions. For example, the adjacency function can be improved and so extended to the use of four parameters, which would allow us to modify its behavior. Figure 4.3 illustrates uses of these four parameters: the geometry *geom* of reference SF, the overlap distance with the reference SF *d1*, the scope *d2* of the relation and the nature of the phenomenon to observe ("neighborhood", "pass", "hut", "fall", etc.). Figure 4.3(a) corresponds to the definition of a geographic zone that, at the same time, covers (over a distance d1) a part of the reference SF and extends it (over a distance d2). Figures 4.3(b) and (c) illustrate the restriction of the scope of the zone thus defined to the zones of a given category (e.g. "forest" and "commune" – (b) and (c), respectively).

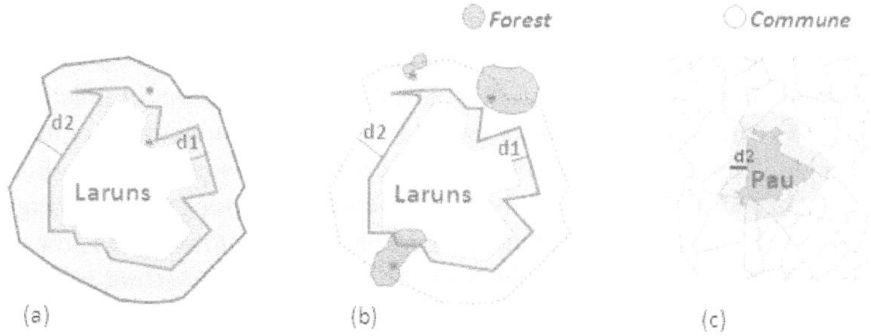

Figure 4.3. *Interpretation of adjacency: advanced adjacency function*

It would be equally interesting to rework the punctual or linear SFs. The representation of an adjacency relative to a peak or to a thalweg line could therefore be adjusted as a function of an interval whose bounds correspond to elevation levels. Figure 4.4 presents the valley of Aspe by visualizing the corresponding thalweg line combined with an altitude limitation of 1,200 m.

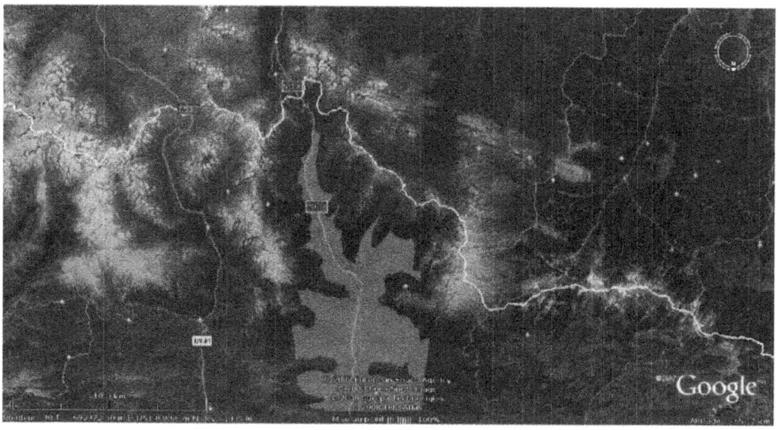

Figure 4.4. *Polygon of a possible representation of the valley of Aspe and corresponding thalweg line traced in Google Earth*

According to principles similar to the processing of adjacency, we propose improving the interpretation of the orientation relation currently supported by bounding box-type representations. Figure 4.5 illustrates a representation of the orientation function. This functions contains five parameters: the geometry *geom* of the reference SF, the direction D_{ir} (here *south*) indicated, the distance D_{is} corresponding to the scope of the orientation relation, the angle β of construction of the form of the representation and the context of implication of the reference SF (*in*, *out* or *inOut*).

Scharl and Tochtermann [SCH 07] qualify spatial relations of "vague", due to the difficulty of modeling subjective natural knowledge. Thus, the fuzzy notion seems equally interesting to us in the context of the approximation of relative spatial features (RSFs) by numeric representations. We propose dividing a representation into geographic zones associated with levels of reliability as illustrated in Figure 4.6. The most relevant zone has a level of reliability of n (3, for example, in Figure 4.6) and the least relevant zone corresponds to a level that tends toward 0 (1, for example, in Figure 4.6). This approach allows us to integrate a level of uncertainty in the indexing phase. In the IR phase, the spatial areas represented as such will be compared to those of the query by the matching function; the reliability ratings will have an impact on the calculated relevance score. So, this fuzzy notion, transcribed in the geometries relative to the indexed features, will allow us to extend a query with a small number of results or restrict a large set of results to the most relevant ones.

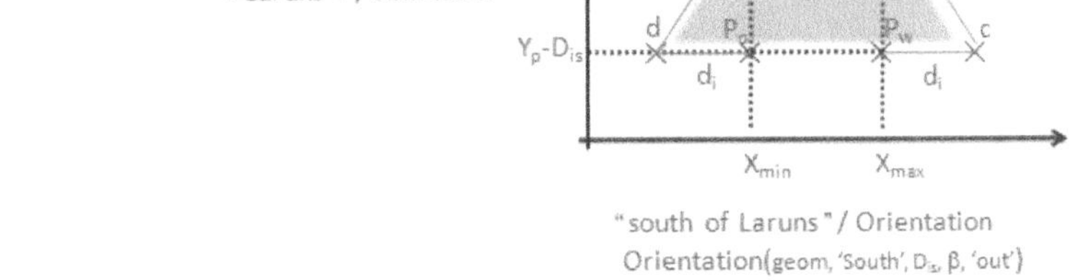

Figure 4.5. *Interpretation of orientation: orientation function*

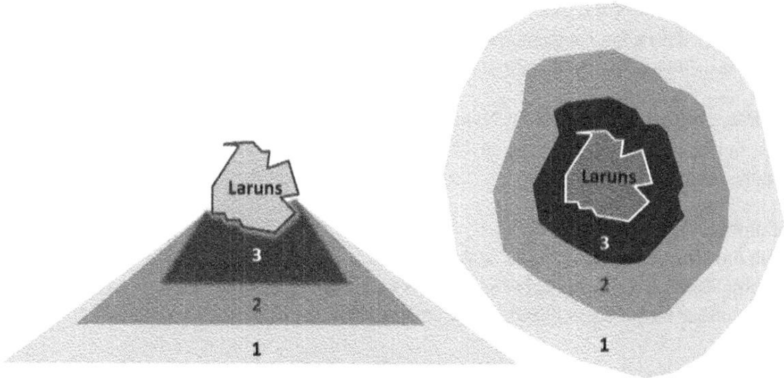

Figure 4.6. *Interpretation of orientation and adjacency with different levels of uncertainty*

An experiment [SAL 08] has allowed us to show an improvement in the quality of GIR results linked to the calculation of more accurate geometries for the representation of the spatial relations of adjacency and inclusion. These new algorithms target the improvement of the precision of numeric representations of RSFs and reduce the problem of noise linked to the approximations by bounding boxes. It would be interesting to continue exploring this path, integrating the notion of "vagueness".

4.2.2. *Interdimensional axis*

We have developed a GIR system combining spatial, temporal and thematic IRSs. The scores of the documents returned by this GIRS mainly depend on those obtained for each of the dimensions involved in the query. We now propose the development of a targeted semantic analysis of the retrieved documents. Once a document is pronounced relevant, n pieces of information (spatial, temporal and/or thematic features) of the document match all or a part of the search criteria expressed in the query. We propose analyzing the relations between these relevant features in the document. This will be followed by a new weighing of this document. This weighing will be based on reinforcing or opposing aspects of those relevant features. The list of proposed results is thus re-ranked. The first stage of the GIR proceeds as a large sieve to propose a first list of documents. The second, more accurate stage of analysis, because it targets the study of relations between features pronounced relevant, would allow us to review the result list.

Figure 4.7 illustrates three types of results obtained for the same query. According to the case, an analysis of the parts that have allowed us to qualify the document in the result list can allow us to recalculate the relevance level of the document.

1. Case of reinforcement of the global relevance
   a) " He finished his hiking in Pagolle at the inn of Platane from where a txistu tune was coming."
   b) " He finished his hiking in Pagolle at the inn of Platane. A txistu tune was coming from inside."

2. Case of downgrading of the global relevance
   a) " He finished his hiking in Pagolle at the inn of Platane. Unfortunately, that day, no txistu tune brightened our sprits."

3. Case of ambiguity
   a) " He finished his hiking in Pagolle at the inn of Platane, a txistu tune was coming from the city hall."
   b) " He finished his hiking in Pagolle at the inn of Platane. A txistu tune was coming from the city hall."
   c) " He finished his hiking in Pagolle at the inn of Platane, a txistu tune was coming from the neighboring village."
   d) " He finished his hiking in Pagolle at the inn of Platane. A txistu tune was coming from the neighboring village."

Figure 4.7. *Example of a search targeting documents invoking "the inn of Platane in the commune of Pagolle and the musical instruments" – "txistu" is a flute defined in the corresponding musical domain ontology*

This *a posteriori* analysis is justified for several reasons:

– The in-depth analysis, *a priori*, of the document repository requires too much significant resources; the GIR system that we have proposed is based on targeted and light natural language processing (NLP) tools as well as IR tools to create a first result list.

– The sentences to be analyzed as well as the relevant spatial, temporal and thematic features within these sentences are only known after the execution of the query; this results in a new targeted analysis of the features and of their relationships performed *a posteriori*.

The information to be analyzed can be found in the same sentence or in different sentences (Figure 4.7 cases 1.a, 3.a and 3.c as well as cases 1.b, 2.a, 3.b and 3.d, respectively). The tools used in the analysis will consequently have to be adapted to the analysis of elements located within a single sentence or a series of sentences, which we call the discourse.

The working hypothesis relative to the two following sections is therefore that the existence of a semantic relation between the features involved in the selection of the document reinforces its relevance. The score of a document now reflects its global relevance with respect to the query rather than the weighted sum of the scores corresponding to each of the search criteria evaluated independently.

4.2.2.1. *Relations in the sentence*

We are interested in the relations between relevant phrases for a given IR need. In this context, there exists a proposition of the W3C aiming at the modeling of relations between phrases composing a sentence.

– N-ary relations.

The W3C[3] presents n-ary relations according to four use cases. The resource description framework/RDF schema (RDF/RDFS) or web ontology language (OWL) languages being devoted to the representation of binary relations, the main objective of this recommendation is to issue propositions adapted to the representation of relations of degrees higher than two.

Four use cases are thus proposed by the W3C. Cases 1 and 2 correspond to forms of extension of binary relations. Cases 3 and 4 describe the establishment of a relationship between several arguments considered to be equal in their participation in the modeled relation. We illustrate these four cases very briefly as follows:

1) "Marco reached the inn of Platane by walking until Pagolle" illustrates a relation between a person, an object (place) and an additional piece of information detailing the entirety of this relation.

2) "Marco reached the inn of Platane from where a txistu tune was coming" illustrates a relation between a person, an object (place) and an additional piece of information relative to this object.

3) "Marco plays the txistu at the inn of Platane, after the mid-day meal" illustrates a relation between a person and several independent arguments, which, generally, give it a meaning: *who does what, where, when, how and why?* Grammatically, there is no difference with the first case. Semantically, the meaning of the arguments allows us to differentiate the two cases.

4) "Marco visited the village of Pagolle, then the pass of Osquich and finally, the chapel of Saint-Antoine" illustrates a relation between a person and a list of arguments (places); the chronology of the enumeration is important.

Figure 4.8 proposes a simplified representation of these four cases of n-ary relations of the W3C. Thus, in the previous examples, case 1 corresponds to a relation containing one or more arguments, these arguments corresponding to the additional pieces of information describing the relation between the subject and the object; case 2 corresponds to a relation containing one or more argument describing the object of the initial relation; case 3 corresponds to a relation containing several connected but independent arguments; finally, case 4 corresponds to a relation

3 http://www.w3.org/TR/swbp-n-aryRelations/.

containing a list of arguments whose semantics are linked to the order of enumeration. Let us note that the most frequently encountered arguments are complements of direct objects, complements of indirect objects and circumstantial complements of time or space.

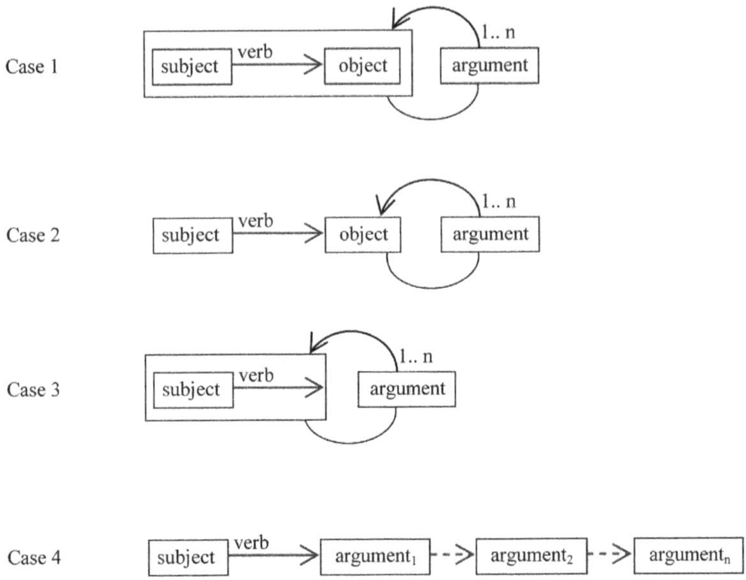

Figure 4.8. *Simplified representation of the four cases of n-ary relations of the W3C*

– N-ary relations and spatial, temporal and thematic information

Let us take an information need that targets the commune of *Pagolle* and the *musical instruments*. Let us now consider a document in which the sentence "He finished his hiking in Pagolle at the inn of Platane from where a txistu tune was coming" has enabled its matching with the spatial and thematic criteria of the query. We can see that, as very frequently, this sentence may be associated with several combinations of W3C cases. Indeed, the first part of the sentence "He finished his hiking in Pagolle at the inn of Platane" arises from case no. 3, whereas the complete sentence arises from case no. 2 because "from where a txistu tune was coming" is an additional piece of information, which is applied to the object "inn of Platane" used as an argument in the first part of the sentence.

This example of case no. 2 illustrates well a relation between the thematic information present in the last argument of the sentence and the spatial information of the first arguments. There is indeed a reinforcement of the relevance of this sentence for the query because a thematic argument and a spatial argument are directly associated via n-ary relation no. 2.

Figure 4.9 presents, in simplified way, the result of the analysis of this sentence by the ALPAGE[4] analyzer, which shows well the explicit relation between "from where a txistu tune was coming" and *"inn of Platane"*.

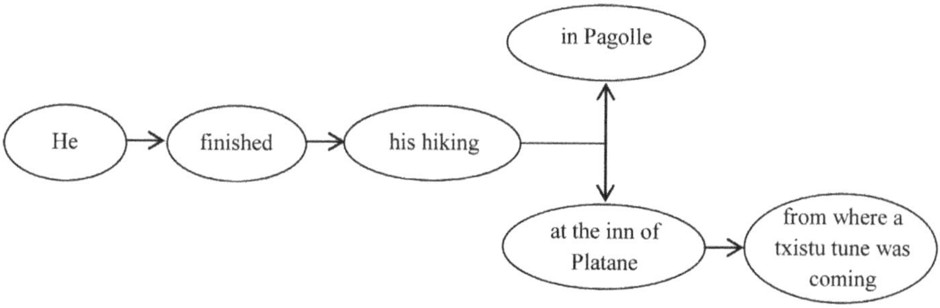

Figure 4.9. *Simplified output of the ALPAGE analyzer*

We issue the hypothesis that spatial, temporal and thematic information associated via a no. 2 n-ary relation (reinforcement of an argument relative to the action by another argument) or a no.3 n-ary relation (association of several arguments via the action they qualify, respectively) reinforce the relevance of the sentence in the framework of the corresponding search.

However, cases 1 ("He left for Pagolle playing a txistu tune") and 4 ("He left Pagolle, walked long hours and finished his itinerary by playing a txistu tune") can imply spatial, temporal or thematic information whose direct relations are not systematic. There is, therefore, no reinforcement to implement here. Finally, different forms of negation should contribute to a downgrading of the document in the set of results or even its pure and simple disqualification.

Numerous approaches are now interested in the semantics of relations. The SemEval[5] campaigns propose, for example, a task called "(8) Multi-Way Classification of Semantic Relations Between Pairs of Nominals", which targets the analysis of relations between two nominal groups in the same sentence. The work of the T2I team of the laboratory of LIUPPA relative to the study of motion or perception verbs [NGU 11] will also be able to contribute to a more accurate analysis of the relations between pieces of spatial information or between pieces of spatial, temporal and thematic information, within the same sentence.

4 http://alpage.inria.fr/alpes/parser.pl.

5 http://www.senseval.org/.

4.2.2.2. *Relations in the discourse*

As before, the relations between spatial, temporal and/or thematic references annotated in a text have to be pointed out. The objective is to review a set of documents pronounced relevant with respect to one or more geographic criteria mentioned in the query.

First, we focus on the relations between the temporal component and the spatial component of the information contained within a document. An annotation model of the beginning and end of scope of a spatial or temporal frame in the text has to be developed as well as a corresponding mechanism of analysis and annotation of such textual sections. We propose linking the semantics to the stream of text [BAT 11] in the goal of attaching a piece of thematic information to a spatial area and/or to a calendar period.

The propositions considered would be based on the work relative to the theory of Charolles [CHA 97, CHA 05b] picked up in recent contributions [BIL 06a, BAT 11]. An approach targeting the identification of discourse frames must therefore be developed; considering the temporal or spatial adverb phrases in front of nouns as the starting point of, respectively, a temporal or spatial frame. The scope of such frames can be extended beyond the sentence introduced by these adverb phrases to the beginning of the next frame, for example. The construction of a frame consists of identifying and tagging the temporal (respectively spatial) introducers followed by evaluating its scope and defining a temporal (respectively spatial) universe for every delimited textual section. This process necessitates the completion of the information available in the spatial and temporal indexes with the study of the structure of the text as well as the identification of the different linguistic tags of transition and connection between the parts of the text.

For a retrieved document, the analysis of its spatial and temporal frames as well as the spatial, temporal and thematic features having served for the selection of the document has to allow us to increase or decrease the corresponding relevance score. The simplest case concerns a document whose relevant SFs are located in a single spatial frame s and the TFs in a single temporal frame t. In the most frequent cases of our corpus, a temporal frame encompasses one or more spatial frames. Some cases can be described following the relations presented in Figures 2.1 (Allen relations [ALL 84]) and 2.2 (RCC-8 [RAN 92] model). These range from the non-connection of a frame s and a frame t to being equal through the partial overlapping of these frames, from the viewpoint of their scope within the text. We thus propose a score-reinforcement coefficient for a document, which is negative if the frames s and t are disjoint, weak if the overlap is partial and significant if there is complete overlapping.

4.2.3. *Expansion of the vocabulary for a qualitative representation of the geographic dimensions*

We are inspired by the *Geooreka!* system described in [BUS 09b]. This system is devoted to the spatial and thematic IR on the Web. Buscaldi and Rosso [BUS 09b] proposed representing a spatial search criteria by a set of toponymic names, which they combine with the other terms of the search and submit to the *yahoo!* engine.

Following the same principle, we hypothesize that this approach could be extended to the spatial, temporal and thematic criteria. Figure 4.10 describes the main steps relative to the process of user need definition. The spatial criteria are described graphically on a map. The denoted geographic zone serves as a parameter in the retrieval of corresponding toponyms. One or more gazetteers are queried in order to extract the toponyms relative to the zoom level and the geographic zone defined in the previous step. For a geometry covering the commune of Laruns, for example, a knowledge base such as a gazetteer would give us all the oronyms, hydronyms as well as points of interest such as huts, mills and fortifications. A new filtering of the obtained toponyms as a function of the user's profile or other arguments of the query can be considered.

The temporal criteria are also described graphically on a time line. The period framed in such a way serves as a parameter in the retrieval of dates, hours, praxonyms (historical events, diseases, cultural events) or phenonyms (hurricanes, high/low pressure zones, stars, comets), for example. One or more gazetteers are queried in order to extract the terms relative to the zoom level and the period defined in the previous step. For a period covering the year 1804, for example, a gazetteer would give us the year 1804 but also *the Coronation of Napoleon*, *the Empire*, *the Civil Code*, etc. A new filtering of the obtained terms depending on the user's profile or other arguments of the query can be considered.

This approach, which is very pragmatic, can be especially interesting in such contexts of use for which space, time and theme are well bounded and described in gazetteers that may be used in the phase of definition of the user need. For a sports reporter specialized in rugby, for example, the spatial gazetteer could be used for the description of the different playing areas of rugby terrain, a temporal gazetteer would describe the periods of game relative to a match and finally, a thematic gazetteer would be devoted to the rugby-specific vocabulary. Likewise, in another framework of use, the spatial gazetteer could describe the rugby world (hemispheres, countries, provinces, clubs, etc.) and the temporal gazetteer could describe the dates and events that made rugby history, for example.

We can thus imagine ad hoc gazetteers describing a building, a set of buildings, opening times, manufacturing times, sale periods and specific domain vocabularies. The *qualitative* representation of the initial version of the query is then submitted to a classic search engine.

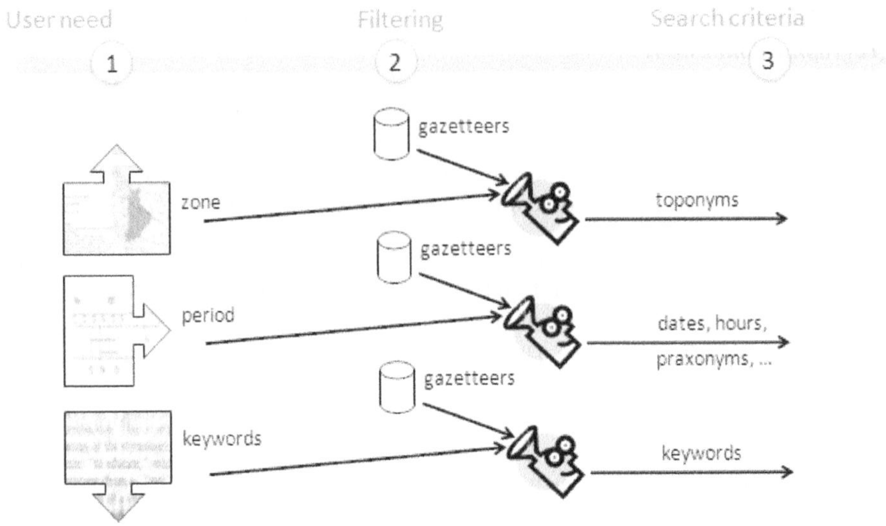

Figure 4.10. *Qualitative translation of a need defined by spatial, temporal and thematic criteria*

Bibliography

[ABO 03] ABOLHASSANI M., FUHR N., GÖVERT N., "Information extraction and automatic markup for XML documents", in BLANKEN H.M., GRABS T., SCHEK H.-J., SCHENKEL R., WEIKUM G. (eds), *Intelligent Search on XML Data*, Lecture Notes in Computer Science, vol. 2818, Springer, pp. 159–178, 2003.

[AGI 07] AGIRRE E., DE LACALLE O.L., MAGNINI B., *et al.*, "SemEval-2007 task 01: evaluating word sense disambiguation (WSD) on cross-language information retrieval", in PETERS C., JIJKOUN V., MANDL T., MÜLLER H., OARD D.W., PEÑAS A., PETRAS V., SANTOS D. (eds), *Cross Language Evaluation forum* (*CLEF*), Lecture Notes in Computer Science, vol. 5152, Springer, pp. 908–917, 2007.

[ALL 84] ALLEN J., "Towards a general theory of action and time", *Artificial Intelligence*, vol. 23, no. 2, pp. 123–154, 1984.

[ALL 91] ALLEN J.F., FIKES R., SANDEWALL E., "Principles of knowledge representation and reasoning", *Proceedings of the 2nd International Conference on Principles of Knowledge Representation and Reasoning*, Morgan Kaufmann, (KR'91), Cambridge, MA, USA, April 22–25, 1991.

[ALO 08] ALONSO O.R., Temporal information retrieval, PhD Thesis, University of California, Davis, CA, 2008.

[ALO 11] ALONSO O., STRÖTGEN J., BAEZA-YATES R., *et al.*, "Temporal information retrieval: challenges and opportunities", *Proceedings of the 1st International Temporal Web Analytics Workshop (TWAW 2011)*, Hyderabad, India, pp. 1–8, 2011.

[AND 10] ANDOGAH G., Geographically constrained information retrieval, PhD Thesis, University of Groningen, The Netherlands, May 2010.

[ARS 85] ARSAC J. (ed.), *Proceedings of the 1st International Conference on Computer-Assisted Information Retrieval, RIAO*, University of Grenoble, France, CID, 1985.

[AUL 10] AULA A., KHAN R.M., GUAN Z., "How does search behavior change as search becomes more difficult?", *CHI '10: Proceedings of the 28th International Conference on Human Factors in Computing Systems*, ACM Press, New York, pp. 35–44, 2010.

[AYA 06] AYACHE S., GENSEL J., QUÉNOT G., "CLIPS-LSR experiments at TRECVID 2006", *TREC Workshop on Video Retrieval Evaluation*, Gaithersburg, USA, 2006.

[BAE 99] BAEZA-YATES R., RIBEIRO-NETO B., *Modern Information Retrieval*, ACM Press, New York, 1999.

[BAS 12] BAST H., BÄURLE F., BUCHHOLD B., *et al.*, "Broccoli: semantic full-text search at your fingertips", Computing Research Repository (*CoRR*), vol. abs/1207.2615, 2012.

[BAT 11] BATTISTELLI D., *Linguistique et recherche d'information, la problématique du temps*, Collection Traitement de l'information, Hermes-Lavoisier, 2011.

[BAU 85] BAUER G., *Namenkunde des Deutschen*, Germanistische, vol. 21, p. 356, 1985.

[BAZ 05] BAZIZ M., Indexation conceptuelle guidée par ontologie pour la recherche d'information, PhD Thesis, IRIT, Paul Sabatier University, Toulouse, France, 2005.

[BAZ 07] BAZIZ M., BOUGHANEM M., PASI G., *et al.*, "An information retrieval driven by ontology: from query to document expansion", in EVANS D., FURUI S., SOULÉDUPUY C. (eds), *Conference on Computer-assisted Information Retrieval (RIAO)*, Centre d'Information Documentaire (CID), 2007.

[BEA 97] BEARD K., SHARMA V., "Multidimensional ranking for data in digital spatial libraries", *International Journal of Digital Libraries*, vol. 1, no. 2, pp. 153–160, 1997.

[BES 04] BESANÇON R., "Technologies statistiques pour la recherche d'informations: les modèles vectoriels", *Les systèmes de recherche d'informations*, Chapter 2, Hermes-Lavoisier, pp. 35–54, 2004.

[BIL 03] BILHAUT F., CHARNOIS T., ENJALBERT P., *et al.*, "Geographic reference analysis for geographic document querying", *HLT-NAACL'03: Proceedings of the Workshop on Analysis of Geographic References*, ACL, Morristown, NJ, pp. 55–62, 2003.

[BIL 06a] BILHAUT F., Analyse automatique de structures thématiques discursives, Application à la recherche d'information, PhD Thesis, University of Caen Basse-Normandie, France, 2006.

[BIL 06b] BILHAUT F., WIDLÖCHER A., "LinguaStream: an integrated environment for computational linguistics experimentation", *EACL'06: Proceedings of the 11th Conference of the European Chapter of the Association for Computational Linguistics: Posters and Demonstrations*, ACL, Morristown, NJ, pp. 95–98, 2006.

[BIL 07] BILHAUT F., DUMONCEL F., ENJALBERT P., *et al.*, "Indexation sémantique et recherche d'information interactive", *CORIA'07: Actes de la 4e Conférence en Recherche d'Information et Applications*, University of Saint-Etienne, pp. 65–76, 2007.

[BOR 98] BORILLO A., *L'espace et son expression en français*, L'essentiel, Ophrys, 1998.

[BOU 93] BOUYSSOU D., PERNY P., PIRLOT M., *et al.*, "A manifesto for the new MCDM era", *Journal of Multi-Criteria Decision Analysis*, vol. 2, pp. 125–127, 1993.

[BOU 08] BOUGHANEM M., SAVOY J. (eds), *Recherche d'information, états des lieux et perspectives*, Hermes-Lavoisier, April 2008.

[BOU 09] BOUAMOR H., "Extraction des connaissances à partir du web pour la recherche des images géoréférencées", *Conférence en Recherche Information et Applications (CORIA)*, Laboratoire des Sciences de l'Information et des Systèmes – Université du Sud Toulon-Var (LSIS-USTV), pp. 519–526, 2009.

[BRI 10] BRISABOA N., LUACES M., PLACES A., *et al.*, "Exploiting geographic references of documents in a geographical information retrieval system using an ontology-based index", *GeoInformatica*, vol. 14, no. 3, pp. 307–331, 2010.

[BUC 00] BUCKLEY C., VOORHEES E.M., "Evaluating evaluation measure stability", *SIGIR'00: Proceedings of the 23rd International ACM SIGIR Conference*, ACM Press, New York, NY, pp. 33–40, 2000.

[BUC 05] BUCHER B., CLOUGH P., JOHO H., *et al.*, "Geographic IR systems: requirements and evaluation", *ICC'05: Proceedings of the 22nd International Cartographic Conference*, Sociedad Española de Cartografía Fotogrametría y Teledetección, 2005.

[BUS 08] BUSCALDI D., ROSSO P., "Using GeoWordNet for geographical information retrieval", in PETERS C., DESELAERS T., FERRO N., GONZALO J., JONES G.J.F., KURIMO M., MANDL T., PEÑAS A., PETRAS V. (eds), *CLEF*, Lecture Notes in Computer Science, vol. 5706, Springer, pp. 863–866, 2008.

[BUS 09a] BUSCALDI D., "Toponym ambiguity in geographical information retrieval", *Proceedings of the 32nd international ACM SIGIR Conference on Research and Development in Information Retrieval, SIGIR '09*, ACM Press, New York, NY, pp. 847–847, 2009.

[BUS 09b] BUSCALDI D., ROSSO P., "Geooreka: enhancing web searches with geographical information", in ANTONELLIS V.D., CASTANO S., CATANIA B., GUERRINI G. (eds), *SEBD'09: Proceedings of the 17th Italian Symposium on Advanced Database Systems*, Edizioni Seneca, pp. 205–212, 2009.

[BÜY 10] BÜYÜKÖZKAN G., RUAN D., "Choquet integral based aggregation approach to software development risk assessment", *Information Sciences*, vol. 180, no. 3, pp. 441–451, 2010.

[CAB 11] CABANAC G., PALACIO D., SALLABERRY C., *et al.*, "Évaluation de la pertinence des résultats en recherche d'information géographique: définition d'un cadre expérimental et validation de l'apport des dimensions de l'information géographique", *Document Numérique*, vol. 14, pp. 169–191, 2011.

[CAR 11] CARDOSO N., "Evaluating geographic information retrieval", *SIGSPATIAL Special*, vol. 3, no. 2, pp. 46–53, 2011.

[CHA 97] CHAROLLES M., "L'encadrement du discours: univers, champs, domaines et Espaces", *Cahier de Recherche Linguistique*, vol. 6 URA-CNRS 1035, Université Nancy 2, pp. 1–73, 1997.

[CHA 05a] CHALLITA K., Problèmes de satisfaction de contraintes spatiales: de l'algèbre des régions à la géométrie affine, PhD Thesis, University of Toulouse III, 2005.

[CHA 05b] CHAROLLES M., "Framing adverbials and their role in discourse cohesion from connexion to forward labelling", *Symposium on the Exploration and Modelling of Meaning (SEM'05)*, pp. 1–18, 2005.

[CHE 07] CHEN Y.-F.R., DI FABBRIZIO G., GIBBON D., *et al.*, "Geotracker: geospatial and temporal RSS navigation", *WWW'07: Proceedings of the 16th International Conference on World Wide Web*, ACM Press, New York, pp. 41–50, 2007.

[CHI 97] CHINCHOR N., "MUC-7 named entity task definition (version 3.5)", *Proceedings of the 7th Message Understanding Conference (MUC-7)*, 1997.

[CHI 98] CHINCHOR N., ROBINSON P., "MUC-7 named entity task definition (version 3.5)", *Proceedings of the 7th Message Understanding Conference (MUC-7)*, 1998.

[CHR 90] CHRISMAN N.R., "Deficiencies of sheets and tiles: building sheetless databases", *International Journal of Geographical Information Science*, vol. 4, no. 2, pp. 157–167, 1990.

[CLE 63] CLEVERDON C., "Comparative efficiency of indexing systems", *Cranfield, Nature*, vol. 197, no. 4863, pp. 129–130, 1963.

[CLO 05] CLOUGH P., "Extracting metadata for spatially-aware information retrieval on the Internet", *Proceedings of the 2005 Workshop on Geographic Information Retrieval (GIR'05)*, ACM Press, New York, pp. 25–30, 2005.

[CLO 06] CLOUGH P., JOHO H., PURVES R., "Judging the spatial relevance of documents for GIR", *ECIR'06: Proceedings of the 28th European Conference on IR Research*, Lecture Notes in Computer Science, vol. 3936, Springer, pp. 548–552, 2006.

[COH 96] COHN A.G., "Calculi for qualitative spatial reasoning", *AISMC-3: Proceedings of the International Conference AISMC-3 on Artificial Intelligence and Symbolic Mathematical Computation*, Springer-Verlag, London, pp. 124–143, 1996.

[COH 01] COHN A.G., HAZARIKA S.M., "Qualitative spatial representation and reasoning: an overview", *Fundamenta Informaticae*, vol. 46, no. 1–2, pp. 1–29, 2001.

[CRO 00] CROFT B., HARABAGIU S., *Advances in Information Retrieval: Recent Research from the Center for Intelligent Information Retrieval*, Kluwer Academic Publishers, 2000.

[CUN 02] CUNNINGHAM H., MAYNARD D., BONTCHEVA K., *et al.*, "GATE: an architecture for development of robust HLT applications", *ACL'02: Proceedings of the 40th Annual Meeting on Association for Computational Linguistics*, ACL, Morristown, pp. 168–175, 2002.

[DA 09] DA COSTA PEREIRA C., DRAGONI M., PASI G., "Multidimensional relevance: a new aggregation criterion", *ECIR'09: Proceedings of the 31st European Conference on IR Research on Advances in Information Retrieval*, Springer-Verlag, Berlin, Heidelberg, pp. 264–275, 2009.

[DA 12] DA COSTA PEREIRA C., DRAGONI M., PASI G., "Multidimensional relevance: prioritized aggregation in a personalized information retrieval setting", *Information Processing and Management*, vol. 48, no. 2, pp. 340–357, 2012.

[DAO 11] DAOUD M., TAMINE L., BOUGHANEM M., "A personalized search using a semantic distance measure in a graph-based ranking model", *Journal of Information Science*, vol. 37, no. 6, pp. 614–636, 2011.

[DEB 81] DE BORDA J.-C., "Mémoire sur les élections au scrutin", *Histoire de l'Académie Royale des Sciences*, Académie Royale des Sciences, Paris, 1781.

[DEL 04] DE LOUPY C., CRESTAN E., "Traitement automatique des langues et systèmes de recherche d'information", *Les systèmes de recherche d'informations*, Chapter 3, Hermes-Lavoisier, pp. 139–158, 2004.

[DEN 97] DENIS M., "The description of routes: a cognitive approach to the production of spatial discourse", *Cahiers de Psychologie Cognitive*, vol. 16, no. 4, pp. 409–458, 1997.

[DUD 10] DUDOGNON D., HUBERT G., RALALASON B.J.V., "ProxiGénéa: une mesure de similarité conceptuelle", *Colloque Veille Stratégique Scientifique et Technologique (VSST)*, Paul Sabatier University-Toulouse, 25–29 October 2010. Available at http://www.ups-tlse.fr, 2010.

[EAS 03] EASTMAN C.M., JANSEN B.J., "Coverage, relevance, and ranking: the impact of query operators on web search engine results", *ACM Transactions on Information Systems*, vol. 21, no. 4, pp. 383–411, October 2003.

[EGE 94] EGENHOFER M.J., MARK D.M., HERRING J., The 9-intersection: formalism and its use for natural-language spatial predicates, Report, National Center for Geographic Information and Analysis (NCGIA), 1994.

[EGO 11] EGOZI O., MARKOVITCH S., GABRILOVICH E., "Concept-based information retrieval using explicit semantic analysis", *ACM Transactions on Information Systems*, vol. 29, no. 2, p. 8, 2011.

[EZZ 11] EZZAT M., POIBEAU T., "A new scheme for annotating semantic relations between named entities in corpora", in ANGELOVA G., BONTCHEVA K., MITKOV R., NICOLOV N. (eds), *Recent Advances in Natural Language Processing* (*RANLP*), RANLP 2011 Organising Committee, pp. 275–281, 2011.

[FAR 08] FARAH M., VANDERPOOTEN D., "An outranking approach for information retrieval", *Information Retrieval*, vol. 11, no. 4, pp. 315–334, 2008.

[FAR 04] FARHAT J., GIRARD L., "L'avenir des services de référence des bibliothèques universitaires", *Argus*, vol. 33, no. 2, pp. 23–28, Fall 2004.

[FER 06] FERNÁNDEZ M., VALLET D., CASTELLS P., "Probabilistic score normalization for rank aggregation", *European Conference on IR Research* (*ECIR'06*), pp. 553–556, 2006.

[FER 11] FERNÁNDEZ M., CANTADOR I., LOPEZ V., *et al.*, "Semantically enhanced information retrieval: an ontology-based approach", *Journal of Web Semantics*, vol. 9, no. 4, pp. 434–452, 2011.

[FER 04] FERRUCI D., LALLY A.,"UIMA: an architectural approach to unstructured information processing in the corporate research environment", *Natural Language Engineering*, vol. 10, no. 3–4, pp. 327–348, 2004.

[FOX 93] FOX E.A., SHAW J.A., "Combination of multiple searches", in HARMAN D.K. (ed.), *TREC-1: Proceedings of the First Text Retrieval Conference*, NIST, Gaithersburg, MD, pp. 243–252, February 1993.

[FRE 92] FREKSA C., "Using orientation information for qualitative spatial reasoning", in FRANK A.U., CAMPARI I., FORMENTINI U. (eds), *Spatio-Temporal Reasoning*, Lecture Notes in Computer Science, vol. 639, Springer, pp. 162–178, 1992.

[FU 05] FU G., JONES C.B., ABDELMOTY A.I., "Ontology-based spatial query expansion in information retrieval", in MEERSMAN R., TARI Z., HACID M.-S., MYLOPOULOS J., PERNICI B., BABAOGLU Ö., JACOBSEN H.-A., LOYALL J.P., KIFER M., SPACCAPIETRA S. (eds), *On The Move Conferences (OTM) (2)*, Lecture Notes in Computer Science, vol. 3761, Springer, pp. 1466–1482, 2005.

[GAI 01] GAIO M., Traitements de l'information géographique: représentations et structures, Accreditation to direct research, University of Caen, France, 2001.

[GAI 08] GAIO M., SALLABERRY C., ETCHEVERRY P., *et al.*, "A global process to access documents' contents from a geographical point of view", *Journal of Visual Languages and Computing*, vol. 19, no. 1, pp. 3–23, 2008.

[GAI 03] GAIZAUSKAS R.J., "Recent advances in *computational terminology*", in BOURIGAULT D., JACQUEMIN C., and L'HOMME M-C. (eds), *Computational Linguistics*, vol. 29, no. 2, pp. 328–332, 2003.

[GAL 01] GALTON A., "Space, time, and the representation of geographical reality", *Topoi*, vol. 20, no. 2, pp. 173–187, 2001.

[GAN 08] GAN Q., ATTENBERG J., MARKOWETZ A., *et al.*, "Analysis of geographic queries in a search engine log", *LocWeb'08: Proceedings of the 1st International Workshop on Location and the Web*, ACM Press, New York, pp. 49–56, 2008.

[GAR 09] GARCIA-CUMBRERAS M., PEREA-ORTEGA J.M., GARCIA-VEGA M., *et al.*, "Information retrieval with geographical references. Relevant documents filtering vs. query expansion", *Information Processing and Management*, vol. 45, no. 5, pp. 605–614, 2009.

[GEY 05] GEY F.C., LARSON R.R., SANDERSON M., *et al.*, "GeoCLEF'05: the CLEF 2005 cross-language geographic information retrieval track overview", *CLEF'05: Proceedings of the 5th workshop on Cross-Language Evaluation Forum*, Lecture Notes in Computer Science, vol. 4022, Springer, pp. 908–919, 2005.

[GOK 09] GOKER A., DAVIES J., *Information Retrieval: Searching in the 21st Century*, John Wiley & Sons, 2009.

[GOS 05] GOSPODNETIĆ O., HATCHER E., *Lucene in Action*, Manning Publications, 2005.

[GRI 96] GRISHMAN R., SUNDHEIM B., "MUC-6: a brief history", *Proceedings of the 16th Conference on Computational Linguistics*, Association for Computational Linguistics, Copenhagen, Denmark, vol. 1, pp. 466–471, 1996.

[HAL 10] HALPIN H., HERZIG D.M., MIKA P., *et al.*, "Evaluating ad-hoc object retrieval", *Proceedings of the International Workshop on Evaluation of Semantic Technologies (IWEST '10)*, 9th International Semantic Web Conference (ISWC '10), Shanghai, PR China, November 2010.

[HAM 09] HAMDADOU D., LIBOUREL T., "Couplage approche multicritère et négociation pour l'aide à la décision en aménagement du territoire", *Conférence internationale de Géomatique et Analyse Spatiale* (*SAGEO*), 2009.

[HAO 10] HAO Q., CAI R., WANG C., *et al.*, "Equip tourists with knowledge mined from travelogues", *Proceedings of the 19th International Conference on World Wide Web (WWW'10)*, ACM Press, New York, pp. 401–410, 2010.

[HIL 90] HILL L., Access to geographic concepts in online bibliographic files: effectiveness of current practices and the potential of a graphic interface, PhD Thesis, University of Pittsburgh, PA, May 1990.

[HIL 06] HILL L.L., *Georeferencing: The Geographic Associations of Information, Digital Libraries and Electronic Publishing*, MIT Press, Cambridge, MA, 2006.

[HOF 12] HOFFART J., SUCHANEK F.M., BERBERICH K., *et al.*, "YAGO2: a spatially and temporally enhanced knowledge base from Wikipedia", *Artificial Intelligence Journal*, Special Issue on Artificial Intelligence, Wikipedia and Semi-Structured Resources, vol. 194, pp. 28–61, 2012.

[HUB 11] HUBERT G., CABANAC G., SALLABERRY C., *et al.*, "Query operators shown beneficial for improving search results", in GRADMANN S., BORRI F., MEGHINI C., SCHULDT H. (eds), *TPDL, Lecture Notes in Computer Science*, vol. 6966, Springer, pp. 118–129, 2011.

[JEN 01] JENSEN C.S., SCHNEIDER M., SEEGER B., *et al.*, (eds), *Proceedings of the 1st International Symposium on Advances in Spatial and Temporal Databases, SSTD*, Redondo Beach, CA, Lecture Notes in Computer Science, vol. 2121, Springer, 2001.

[JOL 11] JOLIVEAU T., ULTSCH J., ROYER A., *et al.*, "Toward the spatial and temporal management of documents: the GeoTopia platform", *Cartographica*, vol. 46, no. 3, pp. 160–169, 2011.

[JON 01] JONES C.B., ALANI H., TUDHOPE D., "Geographical information retrieval with ontologies of place", in D.R. MONTELLO (ed.), *Conference on Spatial Information Theory (COSIT'01)*, vol. 2205/2001, Morro Bay, CA, Springer-Verlag Heidelberg, pp. 322–335, 19–23. September 2001.

[JON 08] JONES R., ZHANG W.V., REY B., *et al.*, "Geographic intention and modification in web search", *International Journal of Geographical Information Science*, vol. 22, no. 3, pp. 229–246, 2008.

[KAN 08] KANHABUA N., NØRVÅG K., "Improving temporal language models for determining time of non-timestamped documents", *ECDL'08: Proceedings of the 12th European Conference on Research and Advanced Technology for Digital Libraries*, Springer-Verlag, Berlin, Heidelberg, pp. 358–370, 2008.

[KEL 07] KELLY D., FU X., "Eliciting better information need descriptions from users of information search systems", *Information Processing and Management*, vol. 43, no. 1, pp. 30–46, 2007.

[KER 09] KERGOSIEN E., KAMEL M., SALLABERRY C., *et al.*, "Construction et enrichissement automatique d'ontologie à partir de ressources externes", in BELLATRECHE L., KASSEL G., THIRAN P. (eds), *JFO'09: Actes des 3e Journées Francophones sur les Ontologies*, ACM SIGAPPFR, pp. 11–20, 2009.

[KER 11] KERGOSIEN E., Indexation, recherche et navigation hybride dans des fonds documentaires territorialisés, PhD Thesis, Team T2I, LIUPPA Laboratory, University of Pau and Pays de l'Adour, France, 2011.

[KEV 11] KEVERS L., Accès sémantique aux Bases de Données Documentaires: Techniques symboliques de traitement automatique du langage pour l'indexation thématique et l'extraction d'informations temporelles, PhD Thesis, Center for Natural Language Processing, Catholic University of Louvain, Belgium, 2011.

[KOW 97] KOWALSKI G., *Information Retrieval Systems: Theory and Implementation*, Kluwer Academic Publishers, 1997.

[KOW 07] KOWALSKI J.-M., CLARAMUNT C., ZUCKER A., "Thalassographeïn: representing maritime spaces in ancient Greece", in WINTER S., DUCKHAM M., KULIK L., KUIPERS B. (eds), *Conference On Spatial Information Theory (COSIT)*, Lecture Notes in Computer Science, vol. 4736, Springer, pp. 47–60, 2007.

[LAB 03] LABREUCHE C., GRABISCH M., "The Choquet integral for the aggregation of interval scales in multicriteria decision making", *Fuzzy Sets and Systems*, vol. 137, no. 1, pp. 11–26, 2003.

[LAR 96] LARSON R.R., "Geographic information retrieval and spatial browsing", *GIS and Libraries: Patrons, Maps and Spatial Information*, University of Illinois, pp. 81–124, April 1996.

[LAR 04] LARSON R.R., FRONTIERA P., "Spatial ranking methods for geographic information retrieval (GIR) in digital libraries", *ECDL'04: Proceedings of the 8th European Conference on Digital Libraries*, Lecture Notes in Computer Science, vol. 3232, Springer, pp. 45–56, 2004.

[LAR 09] LARSON R.R., "Geographic information retrieval and digital libraries", *ECDL'09: Proceedings of the 13th European Conference on Digital Libraries*, Lecture Notes in Computer Science, vol. 5714, Springer, pp. 461–464, 2009.

[LAT 12] LATIRI C.C., GHEZAIEL L.B., AHMED M.B., "Proxemic conceptual network based on ontology enrichment for representing documents in IR", in TEN TEIJE A., VÖLKER J., HANDSCHUH S., STUCKENSCHMIDT H., D'AQUIN M., NIKOLOV A., AUSSENAC-GILLES N., HERNANDEZ N. (eds), *International Conference on Knowledge Engineering and Knowledge Management (EKAW)*, Lecture Notes in Computer Science, vol. 7603, Springer, pp. 72–86, 2012.

[LE 04] LE Y., "A feature-based temporal representation and its implementation with object-relational schema for base geographic data in object-based form", *Student Poster Session*, University Consortium for Geographic Information Science (UCGIS) Assembly, 2004.

[LEB 07] LE BER F., LIGOZAT G., PAPINI O. (eds), *Raisonnement sur l'espace et le temps: des modèles aux applications*, Hermes-Lavoisier, February 2007.

[LEG 10] LE GRAND B., AUFAURE M.-A., SOTO M., "Contextual and conceptual information retrieval and navigation on the web", in CHBEIR R., BADR Y., ABRAHAM A., HASSANIEN A.-E. (eds), *Emergent Web Intelligence: Advanced Information Retrieval, Advanced Information and Knowledge Processing*, Springer London, pp. 1–21, 2010.

[LEI 07] LEIDNER J.L., Toponym resolution in text: annotation, evaluation and applications of spatial grouding of place names, PhD Thesis, Institute for Communicating and Collaborative Systems, School of Informatics, University of Edinburgh, Scotland, May 2007.

[LEI 11] LEIDNER J.L., LIEBERMAN M.D., "Detecting geographical references in the form of place names and associated spatial natural language", *SIGSPATIAL Special*, vol. 3, no. 2, pp. 5–11, 2011.

[LEP 07] LE PARC-LACAYRELLE A., GAIO M., SALLABERRY C., "La composante temps dans l'information géographique textuelle", *Document Numérique*, vol. 10, no. 2, pp. 129–148, 2007.

[LES 06] LESBEGUERIES J., GAIO M., LOUSTAU P., *et al.*, "Geographical information access for non-structured data", *SAC'06: Proceedings of the 2006 ACM Symposium on Applied Computing*, ACM Press, New York, NY, pp. 83–89, 2006.

[LES 07] LESBEGUERIES J., Plate-forme pour l'indexation spatiale multi-niveaux d'un corpus territorialisé, PhD Thesis, Team T2I, LIUPPA Laboratory, University of Pau and Pays de l'Adour, France, 2007.

[LI 10] LI D.H., LAURENT A., PONCELET P., *et al.*, "Extraction of unexpected sentences: a sentiment classification assessed approach", *Intelligent Data Analysis* , vol. 14, no. 1, pp. 31–46, 2010.

[LIB 08] LIBERGE L., GERLINGS J., "Cultural heritage on the (geographical) map", *Museums and the Web 2008: Proceedings*, Montréal, Québec, pp. 163–172, 2008.

[LIE 07] LIEBERMAN M.D., SAMET H., SANKARANARAYANAN J., *et al.*, "STEWARD: architecture of a spatio-textual search engine", *GIS'07: Proceedings of the 15th Annual ACM International Symposium on Advances in Geographic Information Systems*, ACM Press, New York, NY, pp. 1–8, 2007.

[LIE 09] LIESAPUTRA V., WITTEN I.H., BAINBRIDGE D., "Searching in a book", *ECDL'09: Proceedings of the 13th European Conference on Digital Libraries*, Lecture Notes in Computer Science, vol. 5714, Springer, pp. 442–446, 2009.

[LIG 98] LIGOZAT G., "Reasoning about cardinal directions", *Journal of Visual Languages and Computing*, vol. 9, no. 1, pp. 23–44, 1998.

[LIN 98] LIN D., "An information-theoretic definition of similarity", *Proceedings of the 15th International Conference on Machine Learning (ICML'98)*, Morgan Kaufman Publishers Inc., San Francisco, CA, pp. 296–304, 1998.

[LIN 06] LIN J., DEMNER-FUSHMAN D., "The role of knowledge in conceptual retrieval: a study in the domain of clinical medicine", *Proceedings of the 29th Annual International ACM SIGIR Conference on Research and Development in Information Retrieval (SIGIR'06)*, ACM Press, New York, NY, pp. 99–106, 2006.

[LIU 06] LIU Z., GIBBON D.C., SHAHRARAY B., "Multimedia content acquisition and processing in the MIRACLE system", *CCNC'06: Proceedings of the 3rd IEEE Conference on Consumer Communications and Networking*, pp. 272–276, January 2006.

[LIU 09] LIU Y., GUO Q.H., WIECZOREK J., *et al.*, "Positioning localities based on spatial assertions", *International Journal of Geographical Information Science*, vol. 23, no. 11, pp. 1471–1501, 2009.

[LIU 10] LIU X., JIAN C., LU C.-T., "A spatio-temporal-textual crime search engine", *GIS '10: Proceedings of the 18th SIGSPATIAL International Conference on Advances in Geographic Information Systems*, ACM Press, New York, NY, pp. 528–529, 2010.

[LON 05] LONGLEY P.A., GOODCHILD M.F., MAGUIRE D.J., *et al.*, *Geographic Information Systems and Science*, John Wiley & Sons, 2005.

[LOU 08a] LOUSTAU P., Interprétation automatique d'itinéraires dans des récits de voyages, PhD Thesis, Team T2I, LIUPPA Laboratory, University of Pau and Pays de l'Adour, France, 2008.

[LOU 08b] LOUSTAU P., NODENOT T., GAIO M., "Spatial decision support in the pedagogical area: processing travel stories to discover itineraries hidden beneath the surface", in BERNARD L., FRIIS-CHRISTENSEN A., PUNDT H. (eds), *Association of Geographic Information Laboratories for Europe (AGILE) Conference*, Lecture Notes in Geoinformation and Cartography, Springer, pp. 359–378, 2008.

[MAL 00] MALANDAIN N., La relation Texte/Image -Essai de modélisation dans un corpus géographique, PhD Thesis, SIMEM Graduate School, University of Caen Basse-Normandie France, 2000.

[MAL 03] MALCZEWSKI J., CHAPMAN T., FLEGEL C., *et al.*, "GIS – multicriteria evaluation with ordered weighted averaging (OWA): case study of developing watershed management strategies", *Environment and Planning A*, vol. 35, pp. 1769–1784, 2003.

[MAN 11] MANDL T., "Evaluating GIR: geography-oriented or user-oriented?", *SIGSPATIAL Special*, vol. 3, no. 2, pp. 42–45, July, 2011.

[MAN 09] MANGUINHAS H., MARTINS B., BORBINHA J.L., *et al.*, "The DIGMAP geo-temporal web gazetteer service", *e-Perimetron: International Web Journal on Sciences and Technologies Affined to History of Cartography and Maps*, vol. 4, pp. 9–24, 2009.

[MAN 03] MANI I., "Recent developments in temporal information extraction", in NICOLOV N., BONTCHEVA K., ANGELOVA G., MITKOV R. (eds), *RANLP, Current Issues in Linguistic Theory (CILT)*, vol. 260, John Benjamins, Amsterdam/Philadelphia, pp. 45–60, 2003.

[MAN 08a] MANI I., HITZEMAN J., RICHER J., *et al.*, "SpatialML: annotation scheme, corpora, and tools", European Language Resources Association (ELRA) (ed.), *Proceedings of the 6th International Language Resources and Evaluation (LREC'08)*, Marrakech, Morocco, 28–30 May 2008.

[MAN 08b] MANNING C.D., RAGHAVAN P., SCHÜTZE H., *Introduction to Information Retrieval*, Cambridge University Press, July 2008.

[MAR 99] MARTEL J.-M., "L'aide multicritère la décision: méthodes et applications", *Actes de la conférence CORS-SCRO*, Windsor, Ontario, pp. 6–16, 1999.

[MAR 05] MARTINS B., SILVA M.J., ANDRADE L., "Indexing and ranking in Geo-IR systems", *GIR'05: Proceedings of the 2005 Workshop on Geographic Information Retrieval*, ACM Press, New York, NY, pp. 31–34, 2005.

[MAR 07] MARTINS B., BORBINHA J., PEDROSA G., *et al.*, "Geographically-aware information retrieval for collections of digitized historical maps", *GIR'07: Proceedings of the 4th ACM Workshop on Geographical Information Retrieval*, ACM Press, New York, NY, pp. 39–42, 2007.

[MAR 08a] MARTINS B., Geographically aware web text mining, PhD Thesis, University of Lisbon, Portugal, May 2008.

[MAR 08b] MARTINS B., MANGUINHAS H., BORBINHA J.L., "Extracting and exploring the geo-temporal semantics of textual resources", *International Conference on Semantic Computing (ICSC)*, IEEE Computer Society, p. 19, 2008.

[MAT 11] MATA F., CLARAMUNT C., "GeoST: geographic, thematic and temporal information retrieval from heterogeneous web data sources", in TANAKA K., FRÖHLICH P., KIM K.-S. (eds), *International Symposium on Web and Wireless Geographical Information Systems (W2GIS)*, Lecture Notes in Computer Science, vol. 6574, Springer, p. 520, 2011.

[MAU 11] MAUREL D., FRIBURGER N., ANTOINE J.-Y., *et al.*, "Cascade de transducteurs pour la reconnaissance des entités nommées (CasEN): a transducer cascade to recognize French named entities", *Traitement automatique de la langue (TAL)*, vol. 52, no. 1, pp. 69–96, 2011.

[MEI 10] MEIJ E., TRIESCHNIGG D., DE RIJKE M., *et al.*, "Conceptual language models for domain-specific retrieval", *Information Processing and Management*, vol. 46, no. 4, pp. 448–469, 2010.

[MIK 99] MIKHEEV A., MOENS M., GROVER C., "Named entity recognition without gazetteers", *Conference of the European Chapter of the Association for Computational Linguistics (EACL)*, Bergen, Norway, pp. 1–8, 1999.

[MIR 07] MIRON A., GENSEL J., VILLANOVA-OLIVIER M., *et al.*, "Relations spatiales qualitatives dans les ontologies géographiques avec Ontologies in Arom Knowledge Representation System (ONTOAST)", *Conférence internationale de Géomatique et Analyse Spatiale (SAGEO)*, 2007.

[MUL 04] MULLER P., TANNIER X., "Annotating and measuring temporal relations in texts", *Proceedings of the 20th International Conference on Computational Linguistics (Coling 04)*, Association for Computational Linguistics, Geneva, Switzerland, pp. 50–56, August 2004.

[NGU 11] NGUYEN VAN T., GAIO M., "Utilisation de la relation 'Verbe-Préposition–Toponyme' pour un inventaire lexical automatique", *Lexis and Grammar Conference (LGC) 2011*, Nicosia, Cyprus, pp. 1–8, October 2011.

[OGI 01] OGILVIE P., CALLAN J.P., "Experiments using the lemur toolkit", *TREC'01: Proceedings of the 9th Text Retrieval Conference*, NIST, Gaithersburg, MD, February 2001.

[OHT 12] OHTA T., PYYSALO S., TSUJII J., *et al.*, "Open-domain anatomical entity mention detection", *Proceedings of the ACL Workshop on Detecting Stucture in Scholarly Discourse (DSSD)*, Jeju Island, Corea, pp. 27–36, 2012.

[OUN 05] OUNIS I., AMATI G., PLACHOURAS V., *et al.*, "Terrier information retrieval platform", *ECIR'05: Proceedings of the 27th European Conference on IR Research*, Lecture Notes in Computer Science, vol. 3408, Springer, pp. 517–519, 2005.

[OVE 09] OVERELL S.E., Geographic information retrieval: classification, disambiguation and modelling, PhD Thesis, Department of Computing, Imperial College, London, May 2009.

[PAL 09] PALACIO D., SALLABERRY C., GAIO M., "Normalizing spatial information to better combine criteria in geographical information retrieval", *ECIR GIIW: 31st European Conference on Information Retrieval, Geographic Information on the Internet Workshop (GIIW)*, Toulouse, France, pp. 37–49, April 2009.

[PAL 10a] PALACIO D., Combinaison de critères par contraintes pour la Recherche d'Information Géographique, PhD Thesis, Team T2I, LIUPPA Laboratory, University of Pau and Pays de l'Adour, France, 2010.

[PAL 10b] PALACIO D., CABANAC G., SALLABERRY C., *et al.*, "Measuring effectiveness of geographic IR systems in digital libraries: evaluation framework and case study", in LALMAS M., JOSE J., RAUBER A., SEBASTIANI F., FROMMHOLZ I. (eds), *ECDL '10: Proceedings of the 14th European Conference on Research and Advanced Technology for Digital Libraries*, Lecture Notes in Computer Science, vol. 6273, Springer, pp. 340–351, September 2010.

[PAL 10c] PALACIO D., CABANAC G., SALLABERRY C., *et al.*, "On the evaluation of geographic information retrieval systems-evaluation framework and case study", *International Journal on Digital Libraries*, vol. 11, no. 2, pp. 91–109, 2010.

[PAL 10d] PALACIO D., SALLABERRY C., GAIO M., "Normalizing spatial information to improve geographical information indexing and retrieval in digital libraries", *ISGIS '10: Proceedings of the Joint International Conference on Theory, Data Handling and Modelling in Geospatial Information Science* – also published by Springer-Verlag in Lecture Notes in Geoinformation and Cartography series, pp. 229–234, 2010.

[PAL 11] PALACIO D., SALLABERRY C., GAIO M., "Generalize geographic information", *AGILE: The 14th AGILE International Conference on Geographic Information Science – Poster Session*, Utrecht, The Netherlands, 2011.

[PAL 12a] PALACIO D., SALLABERRY C., CABANAC G., *et al.*, "Do expressive geographic queries lead to improvement in retrieval effectiveness?", in GENSEL J., JOSSELIN D., VANDENBROUCKE D. (eds), *Bridging the Geographic Information Sciences – International AGILE 2012 Conference*, LNG&C, Springer, Avignon, France, pp. 267–286, 2012.

[PAL 12b] PALACIO D., SALLABERRY C., GAIO M., "Normalizing spatial information to improve geographical information indexing and retrieval in digital libraries", in YEH A.G., SHI W., LEUNG Y., ZHOU C. (eds), *Advances in Spatial Data Handling and GIS*, Lecture Notes in Geoinformation and Cartography, Springer, Berlin, Heidelberg, pp. 65–78, 2012.

[PER 08] PEREA-ORTEGA J.M., GARCÍA-CUMBRERAS M.A., GARCÍA-VEGA M., *et al.*, "Comparing several textual information retrieval systems for the geographical information retrieval task", *NLDB'08: Proceedings of the 13th International Conference on Natural Language and Information Systems*, Springer-Verlag, Berlin, Heidelberg, pp. 142–147, 2008.

[PER 07] PERRY M., SHETH A., BUDAK-ARPINAR I., *et al.*, *Geospatial and Temporal Semantic Analytics*", Idea Group, pp. 1–13, 2007.

[PET 01] PETERS C. (ed.), *CLEF'01: Proceedings of the 2nd Workshop of the Cross-Language Evaluation Forum*, Lecture Notes in Computer Science, vol. 2069, Springer, 2001.

[PFO 09] PFOSER D., EFENTAKIS A., HADZILACOS T., *et al.*, "Providing universal access to history textbooks: a modified GIS case", *W2GIS'09: Proceedings of the 9th International Symposium on Web and Wireless Geographical Information Systems*, Lecture Notes in Computer Science, vol. 5886, pp. 87–102, 2009.

[PHA 07] PHAM T.-T., MAILLOT N.E., LIM J.-H., *et al.*, "Latent semantic fusion model for image retrieval and annotation", *CIKM'07: Proceedings of the 16th ACM Conference on Conference on Information and Knowledge Management*, ACM Press, New York, NY, pp. 439–444, 2007.

[PIS 93] PISSINOU N. (ed.), *Proceedings of the 1st ACM Workshop on Advances in Geographic Information Systems*, ACM Press, Arlington, VA, 1993.

[POI 03] POIBEAU T., *Extraction automatique d'information, du texte brut au web sémantique*, Hermes-Lavoisier, 2003.

[POO 10] POON H., DOMINGOS P., "Unsupervised ontology induction from text", in HAJIC J., CARBERRY S., CLARK S. (eds), *The Association for Computer Linguistics* (*ACL*) The Association for Computer Linguistics, pp. 296–305, 2010.

[POW 97] POWER D.J., "What is a DSS?", *The On-Line Executive Journal for Data-Intensive Decision Support*, vol. 1, no. 3, 1997.

[PUR 04] PURVES R., JONES C.B., "Workshop on geographic information retrieval, SIGIR 2004", *SIGIR Forum*, vol. 38, no. 2, pp. 53–56, 2004.

[PUR 07] PURVES R.S., CLOUGH P., JONES C.B., *et al.*, "The design and implementation of SPIRIT: a spatially aware search engine for information retrieval on the Internet", *International Journal of Geographical Information Science*, vol. 21, no. 7, pp. 717–745, 2007.

[PUR 11] PURVES R., JONES C., "Geographic information retrieval", *SIGSPATIAL Special*, vol. 3, no. 2, pp. 2–4, 2011.

[PUS 03] PUSTEJOVSKY J., CASTAÑO J.M., INGRIA R., *et al.*, "TimeML: robust specification of event and temporal expressions in text", in MAYBURY M.T. (ed.), *New Directions in Question Answering*, AAAI Press, pp. 28–34, 2003.

[PUS 05] PUSTEJOVSKY J., KNIPPEN R., LITTMAN J., *et al.*, "Temporal and event information in natural language text", *Language Resources and Evaluation*, vol. 39, pp. 123–164, 2005.

[PUS 10] PUSTEJOVSKY J., LEE K., BUNT H., *et al.*, "ISO-TimeML: an international standard for semantic annotation", in CALZOLARI N., CHOUKRI K., MAEGAARD B., MARIANI J., ODIJK J., PIPERIDIS S., ROSNER M., TAPIAS D. (eds), *Language Resources and Evaluation Conference* (*LREC*), European Language Resources Association, 2010.

[RAN 92] RANDELL D.A., CUI Z., COHN A., "A spatial logic based on regions and connection", in NEBEL B., RICH C., SWARTOUT W. (eds), *Proceedings of the 3rd International Conference on Principles of Knowledge Representation and Reasoning (KR'92)*, Morgan Kaufman, San Mateo, CA, pp. 165–176, 1992.

[RAS 03] RASOLOFO Y., HAWKING D., SAVOY J., "Result merging strategies for a current news metasearcher", *Information Processing and Management*, vol. 39, no. 4, pp. 581–609, June 2003.

[RES 95] RESNIK P., "Using information content to evaluate semantic similarity in a taxonomy", *International Joint Conference on Artificial Intelligence* (*IJCAI*), pp. 448–453, 1995.

[REY 09] REYMONET A., THOMAS J., AUSSENAC-GILLES N., "Ontology based information retrieval: an application to automotive diagnosis", in NYBERG M., FRISK E., KRISANDER M., ASLUND J. (eds), *International Workshop on Principles of Diagnosis (DX)*, Stockholm, Linköping University, Institute of Technology, pp. 9–14, 14–17, June 2009, (guest speaker). Available at http://www.lith.liu.se.

[ROC 09] ROCHE M., PONCELET P., "Fouille de données d'opinion", *Revue Nouvelles Technologies de l'Information (RNTI)*, vol. E-17, pp. 1–202, 2009.

[SAA 80] SAATY T.L., *The Analytic Hierarchy Process, Planning, Priority Setting, Resource Allocation*, McGraw-Hill, New York, 1980.

[SAG 08] SAGOT B., BOULLIER P., "SxPipe 2: architecture pour le traitement présyntaxique de corpus bruts", *Traitement Automatique des Langues*, vol. 49, no. 2, pp. 155–188, 2008.

[SAL 71] SALTON G., *The SMART Retrieval System: Experiments in Automatic Document Processing*, Prentice-Hall, Englewood Cliffs, NJ, 1971.

[SAL 07a] SALLABERRY C., BAZIZ M., LESBEGUERIES J., *et al.*, "Towards an IE and IR system dealing with spatial information in digital libraries – evaluation case study", in CARDOSO J., CORDEIRO J., FILIPE J. (eds), *ICEIS'07: Proceedings of the 9th International Conference on Enterprise Information Systems*, INSTICC Press, pp. 190–197, 2007.

[SAL 07b] SALLABERRY C., BAZIZ M., LESBEGUERIES J., *et al.*, "Une approche d'extraction et de recherche d'information spatiale dans les documents textuels - évaluation", *4th French Information Retrieval Conference, Conférence en Recherche d'Information et Applications (CORIA)*, University of Saint-Etienne, pp. 53–64, 2007.

[SAL 07c] SALLABERRY C., GAIO M., LESBEGUERIES J., *et al.*, "A semantic approach for geospatial information extraction from unstructured documents", *Advanced Information and Knowledge Processing Series –The Geospatial Web*, Chapter 9, Springer, London, UK, 2007.

[SAL 08] SALLABERRY C., GAIO M., PALACIO D., *et al.*, "Fuzzying GIS topological functions for GIR needs", *Proceeding of the 2nd International Workshop on Geographic Information Retrieval (GIR'08)*, ACM Press, New York, NY, pp. 1–8, 2008.

[SAL 09] SALLABERRY C., ROYER A., LOUSTAU P., *et al.*, "GeoStream: spatial information indexing within textual documents supported by a dynamically parameterized web service", in De NANTES E.C. (eds), *OGRS 2009: International Opensource Geospatial Research Symposium*, Nantes, France, July 2009.

[SAL 11] SALLABERRY C., PALACIO D., GAIO M., "Indexation spatiale et temporelle basée sur un principe de "tuilage": contribution à la recherche d'information géographique dans des documents textuels faiblement structurés", in PASI G., BELLOT P. (eds), *CORIA*, Universitaires d'Avignon, pp. 327–334, 2011.

[SAL 12] SALLABERRY C., Accès par le contenu géographique à des corpus textuels complémentarité des dimensions spatiale, temporelle et thématique dans le processus de recherche d'information, Accreditation to direct research, University of Pau and Pays de l'Adour, France, 2012.

[SAL 75] SALTON G., WONG A., YANG C.S., "A vector space model for automatic indexing", *Communications of the ACM*, vol. 18, no. 11, pp. 613–620, November 1975.

[SAN 04] SANDERSON M., KOHLER J., "Analyzing geographic queries", *SIGIR-GIR'04: Proceedings of the Workshop on Geographic Information Retrieval at SIGIR*, Sheffield, United Kingdom, 2004.

[SAN 10] SANDERSON M., "Test collection based evaluation of information retrieval systems", *Foundations and Trends in Information Retrieval*, vol. 4, no. 4, pp. 247–375, 2010.

[SCH 07] SCHARL A., TOCHTERMANN K., "The geospatial web – advanced information and knowledge", *Advanced Information and Knowledge Processing Series*, Springer, London, UK, 2007.

[SIL 99] SILVERSTEIN C., MARAIS H., HENZINGER M., *et al.*, "Analysis of a very large web search engine query log", *SIGIR Forum*, vol. 33, no. 1, pp. 6–12, September 1999.

[SMI 01] SMITH B., MARK D.M., "Geographical categories: an ontological investigation", *International Journal of Geographical Information Science*, vol. 15, no. 7, pp. 591–612, 2001.

[SPI 01] SPINK A., WOLFRAM D., JANSEN M.B.J., *et al.*, "Searching the web: the public and their queries", *Journal of the American Society for Information Science and Technology*, vol. 52, no. 3, pp. 226–234, February 2001.

[STO 08] STOKES N., LI Y., MOFFAT A., *et al.*, "An empirical study of the effects of NLP components on geographic IR performance", *International Journal of Geographical Information Science*, vol. 22, no. 3, pp. 247–264, January 2008.

[STR 10] STRÖTGEN J., GERTZ M., POPOV P., "Extraction and exploration of spatiotemporal information in documents", *GIR '10: Proceedings of the 6th Workshop on Geographic Information Retrieval*, ACM Press, New York, NY, pp. 16:1–16:8, 2010.

[SZM 98] SZMURLO M., GAIO M., "Extended conceptual neighborhoods", *ISPRS Commission IV Symposium on GIS – between Visions and Applications*, pp. 1–8, 1998.

[TAL 00] TALMY L., "How language structures space", *Toward a Cognitive Semantics, Volume I: Conceptual Structuring Systems*, MIT Press, pp. 177–254, 2000.

[TEI 11] TEISSEIDRE C., BATTISTELLI D., MINEL J.-L., "Recherche d'information et temps linguistique: une heuristique pour calculer la pertinence des expressions calendaires", *TALN'11: Actes de la 18e Conférence sur le Traitement Automatique du Langage Naturel*, Montpellier, France, 2011.

[URE 10] UREN V.S., SABOU M., MOTTA E., *et al.*, "Reflections on five years of evaluating semantic search systems", *International Journal of Metadata, Semantics and Ontologies*, vol. 5, no. 2, pp. 87–98, 2010.

[USE 96] USERY E.L., "A feature-based geographic information system model", *Photogrammetric Engineering and Remote Sensing*, vol. 62, no. 7, pp. 833–838, 1996.

[VAI 05] VAID S., JONES C.B., JOHO H., *et al.*, "Spatio-textual indexing for geographical search on the web", *SSTD'05: Proceedings of the 9th International Symposium on Spatial and Temporal Databases*, Lecture Notes in Computer Science, vol. 3633, Springer, pp. 218–235, 2005.

[VAL 06] VALCARTIER, "GRID – geospatial retrieval of indexed document", Technical report, Military research center, Canada, 2006.

[VAN 86] VANDELOISE C., *L'espace en français*, Éditions du Seuil, Paris, 1986.

[VAN 05] VAN KREVELD M., REINBACHER I., ARAMPATZIS A., *et al.*, "Multi-dimensional scattered ranking methods for geographic information retrieval", *GeoInformatica*, vol. 9, pp. 61–84, 2005.

[VER 09] VERHAGEN M., GAIZAUSKAS R., SCHILDER F., *et al.*, "The TempEval challenge: identifying temporal relations in text", *Language Resources and Evaluation*, vol. 43, no. 2, pp. 161–179, 2009.

[VOG 99] VOGT C.C., COTTRELL G.W., "Fusion via a linear combination of scores", *Information Retrieval*, vol. 1, no. 3, pp. 151–173, 1999.

[VOL 10] VOLK M., BUBENHOFER N., ALTHAUS A., *et al.*, "Challenges in building a multilingual alpine heritage corpus", in CHAIR N.C.C., CHOUKRI K., MAEGAARD B., MARIANI J., ODIJK J., PIPERIDIS S., ROSNER M., TAPIAS D. (eds), *Proceedings of the 7th International Conference on Language Resources and Evaluation (LREC '10)*, European Language Resources Association (ELRA), Valletta, Malta, pp. 19–2, May 2010.

[VOO 99] VOORHEES E.M., HARMAN D.K., "Overview of the seventh text retrieval conference (TREC-8)", *TREC-8: Proceedings of the 8th Text Retrieval Conference*, Gaithersburg, USA, pp. 1–23, 1999.

[VOO 02] VOORHEES E.M., "The philosophy of information retrieval evaluation", in PETERS C., BRASCHLER M., GONZALO J., KLUCK M. (eds), *CLEF'01: Proceedings of the 2nd Workshop of the Cross-Language Evaluation Forum*, Lecture Notes in Computer Science, vol. 2406, Springer, pp. 355–370, 2002.

[VOO 05] VOORHEES E.M., HARMAN D.K., *Text Retrieval Conference (TREC): Experiment and Evaluation in Information Retrieval*, MIT Press, Cambridge, MA, 2005.

[WAL 92] WALKER D., NEWMAN I., MEDYCKYJ-SCOTT D., *et al.*, "A system for identifying datasets for GIS users", *International Journal of Geographical Information Systems*, vol. 6, no. 6, pp. 511–527, 1992.

[WHI 07] WHITE R.W., MORRIS D., "Investigating the querying and browsing behavior of advanced search engine users", *SIGIR'07: Proceedings of the 30th Annual International ACM SIGIR Conference*, ACM Press, New York, NY, pp. 255–262, 2007.

[WID 04] WIDLÖCHER A., FAUROT E., BILHAUT F., "Multimodal indexation of contrastive structures in geographical documents", *RIAO'04: Proceedings of the 1st International Conference on Computer-Assisted Information Retrieval*, pp. 555–570, 2004.

[WID 05] WIDLÖCHER A., BILHAUT F., "La plate-forme LinguaStream: un outil d'exploration linguistique sur corpus", *TALN'05: Actes de la 12e Conférence sur le Traitement Automatique du Langage Naturel*, Dourdan, France, pp. 517–522, 2005.

[WOL 92] WOLPERT D.H., "Stacked generalization", *Neural Networks*, vol. 5, pp. 241–259, 1992.

[WOO 94] WOODRUFF A.G., PLAUNT C., "GIPSY: automated geographic indexing of text documents", *Journal of the American Society for Information Science*, vol. 45, no. 9, pp. 645–655, 1994.

[WU 94] WU Z., PALMER M.S., "Verb semantics and lexical selection", in PUSTEJOVSKY J. (ed.), *Association for Computational Linguistics (ACL)*, Morgan Kaufman Publishers/ACL, pp. 133–138, 1994.

[YAG 88] YAGER R.R., "On ordered weighted averaging aggregation operators in multicriteria decisionmaking", *IEEE Transactions on Systems, Man, and Cybernetics*, vol. 18, no. 1, pp. 183–190, 1988.

[ZEL 82] ZELENY M., *Multiple Criteria Decision Making*, McGraw-Hill, New York, 1982.

[ZGH 08] ZGHAL H.B., AUFAURE M.-A., SOUSSI R., "Towards an on-line semantic information retrieval system based on fuzzy ontologies", *Journal of Digital Information Management*, vol. 6, no. 5, pp. 375–385, 2008.

Index

Printed and bound by CPI Group (UK) Ltd, Croydon, CR0 4YY
13/09/2026
14961078-0008